INDUSTRIAL APPLICATIONS OF LASER DIAGNOSTICS

INDUSTRIAL APPLICATIONS OF LASER DIAGNOSTICS

Yoshihiro Deguchi

CRC Press
Taylor & Francis Group
Boca Raton London New York

CRC Press is an imprint of the
Taylor & Francis Group, an **informa** business

CRC Press
Taylor & Francis Group
6000 Broken Sound Parkway NW, Suite 300
Boca Raton, FL 33487-2742

© 2012 by Taylor & Francis Group, LLC
CRC Press is an imprint of Taylor & Francis Group, an Informa business

First issued in paperback 2019

No claim to original U.S. Government works

ISBN-13: 978-0-367-44592-8 (pbk)
ISBN-13: 978-1-4398-5337-5 (hbk)

Library of Congress Cataloging-in-Publication Data

Deguchi, Yoshihiro.
    Industrial applications of laser diagnostics / Yoshihiro Deguchi.
      p. cm.
    Includes bibliographical references and index.
    ISBN 978-1-4398-5337-5 (hardback)
    1. Lasers--Diagnostic use--Industrial applications. 2. Laser spectroscopy. I. Title.

TA1675.D44 2012
621.36'6--dc23
                                                   2011038755

Visit the Taylor & Francis Web site at
http://www.taylorandfrancis.com

and the CRC Press Web site at
http://www.crcpress.com

Contents

Preface

Recent years have seen tighter regulations of harmful substances such as NO_x, CO, heavy metals, particles, and environmental emissions from cars as well as several types of commercial plant discharges. It is also a major challenge to reduce anthropogenic carbon dioxide emissions. Behind the trend is the fact that increased carbon dioxide in the air is a leading cause of global warming and adversely affects natural ecosystems. Further, the demands for lowering the burdens on the environment continue to grow steadily. It is thus becoming more important to understand emission characteristics to minimize environmental disruption and improve the efficiency of industrial machinery and plant processes. Considering the above situation, it is also equally important to monitor controlling factors in order to improve the operation of industrial machinery and plants.

There are several methods to detect the parameters of these emissions and controls, consisting of well-known "industrial standards." These standard methods are well established and easily accessible, although they are limited in terms of meeting industrial needs because of slow response, low sensitivity, complicated preconcentrations, and similar factors. In contrast, laser diagnostics make it possible to monitor these parameters *in situ* and real-time mode because of their fast response, high sensitivity, and noncontact features. Among the many techniques now being proposed, several methods such as tunable diode laser absorption spectroscopy have been intensively employed to meet practical industrial requirements. However, recent advances in laser diagnostics have not shown substantial progress from a practical application standpoint. Although there have been several factors preventing progress on the implementation of practical uses, such as the cost and vulnerability of laser systems, this is in large part due to misinterpretation of the ability of laser diagnostics among researchers and engineers in the fields of industrial machinery and plants. There are also few books and research papers that have been written for the systems engineers in the above mentioned fields.

As can be expected from the title, *Industrial Applications of Laser Diagnostics*, the main theme of this book is the visualization of laser diagnostics for both researchers specializing in laser diagnostics and engineers in machinery and plants. Detailed theoretical explanations and complex equations, which are usually the main inhibitors for unspecialized readers in laser fields, are avoided in the main text and summarized in appendices or indicated by citations. The book contains easy-to-follow indicators to apply laser diagnostics in industrial applications where such methods can be useful.

In Chapter 1, the guideline for industrial applications of laser diagnostics is shown as introductory information followed by a brief exposition of the main components of laser diagnostics systems. Chapter 2 summarizes application codes of laser diagnostics to industrial systems. Readers will catch the flavor of this book with overall information and strategic direction of laser diagnostics applications to industrial systems. Laser-induced fluorescence is discussed in detail in Chapter 3, laser-induced breakdown spectroscopy in Chapter 4, spontaneous Raman spectroscopy and coherent anti-Stokes Raman spectroscopy (CARS) in Chapter 5, tunable diode laser absorption spectroscopy in Chapter 6, and time-of-flight mass spectrometry in Chapter 7. There are also growing biochemical application fields, and they are briefly introduced in Chapter 8.

About the Author

Yoshihiro Deguchi began his career in laser diagnostics with BE, ME, and DE degrees from Toyohashi University of Technology, Toyohashi, Japan, in 1985, 1987, and 1990. After receiving his DE, he worked as a research engineer in applied physics fields for Mitsubishi Heavy Industries, Ltd. for twenty years. He engaged in developing laser diagnostics such as laser-induced fluorescence, laser-induced breakdown spectroscopy, and laser Raman spectroscopy to apply these techniques to industrial fields. He moved to the University of Tokushima, Tokushima, Japan, as a full professor in 2010. Dr. Deguchi has published research papers in the area of industrial applications of laser diagnostics. He is one of the leading engineers to put laser diagnostics into practical use, especially in large-scale plants.

In addition to research interests, Dr. Deguchi teaches the encouragement and use of intellectual property. Qualified as both a patent attorney and a professional engineer, he works on educational projects that induce student spontaneity regarding their potential inventions at the University of Tokushima.

Units, Symbols, and Acronyms

Symbols and acronyms used in this book

Symbols

A	Einstein A coefficient (1/s)
A_j	Integrated absorbance
B_e	Rotational constant
c	Velocity of light
D_e	Rotational constant
E	Energy (J)
E	Electric field
G	Line-shape function
g	Statistical weight
h	Planck's constant (Js)
I	Light intensity (W/m^2)
K	Proportional constant
k	Boltzmann constant (J/K)
ℓ	Length (m)
m	Mass
n	Number density (1/m^3)
P_2	Predissociation rate (1/s)
p	Oscillating dipole moment
Q	Coordinate
Q_{21}	Quenching rate (1/s)
r	Radius (m)
$S_{i,j}$	Line strength of the transition from i to j states
T	Temperature (K)
t	Time (s)
V	Volume (m^3)
V_p	Electric field potential
v	Velocity (m/s)
W	Stimulated emission or absorption rate (1/s)
x,y,z	Axis

z	Ion value
α	Absorbance
α	Polarizability
α_e	Vibration-rotation interaction constant (cm^{-1})
β_e	Vibration-rotation interaction constant (cm^{-1})
χ_{CARS}	CARS susceptibility
χ_{nr}	Nonresonant susceptibility
χ_r	Resonant susceptibility
χ'	Real part of the resonant susceptibility
χ''	Imaginary part of the resonant susceptibility
ε	Efficiency
Φ	Electric potential
Γ	Constant related to the spontaneous Raman line width
κ	Absorption coefficient
λ	Wavelength
ν	Frequency of light (1/s)
$d\sigma/d\Omega$	Differential scattering cross section (m^2/sr)
Ω	Solid angle (sr)
ω	Angular frequency (1/s)
ω_e	Vibrational constant (cm^{-1})
$\omega_e x_e$	Vibrational constant (cm^{-1})

Acronyms

1-D	One-dimensional
2-D	Two-dimensional
3-D	Three-dimensional
BBO	β-BaB_2O_4
BOTDR	Brillouin optical-time-domain reflectometer
CAD	Crank-angle degree
CARS	Coherent anti-Stokes Raman spectroscopy
CCA	Chromate copper arsenate
CCD	Charge-coupled device
CFD	Computational fluid dynamics
CRDS	Cavity ring-down spectroscopy
CT	Computer tomography

CVD	Chemical vapor deposition
DBR	Distributed Bragg reflector
DFB	Distributed feedback
DMA	Differential mobility analyzer
DPSS	Diode-pumped solid state
DXN	Dioxin
ECDL	External cavity diode laser
EM-CCD	Electron-multiplying CCD
FFT	Fast Fourier transform
FD-OCT	Frequency domain OCT
fs	Femtosecond
FTIR	Fourier transform infrared spectroscopy
FWHM	Full width at half maximum
ICCD	Image-intensified CCD
ICP-AES	Inductively coupled plasma—atomic emission spectroscopy
ICP-MS	Inductively coupled plasma—mass spectroscopy
IMS	Imaging mass spectrometry
KTP	$KTiOPO_4$
LDV	Laser Doppler velocimetry
LIBS	Laser-induced breakdown spectroscopy
LIF	Laser-induced fluorescence
LII	Laser-induced incandescence
LIPS	Laser-induced plasma spectroscopy
LI-TOFMS	Laser ionization time-of-flight mass spectrometry
MALDI	Matrix-assisted laser desorption/ionization
MCB	Monochlorobenzene
MCP	Microchannel plate
MIR	Mid-infrared
NDIR	Nondispersive infrared spectroscopy
Nd:YAG	Neodymium-dopedyttrium aluminium garnet
Nd:YLF	Neodymium-dopedyttrium lithium fluoride
NIR	Near infrared
OCT	Optical coherence tomography
OPO	Optical parametric oscillator
PAH	Polycyclic aromatic hydrocarbon

PAS	Photoacoustic spectroscopy
PCB	Polychlorinated biphenyls
PIV	Particle image velocimetry
PLIF	Planar LIF
PMT	Photomultiplier tube
ppb	Parts per billion (10^{-9})
ppm	Parts per million (10^{-6})
ppt	Parts per trillion (10^{-12})
ps	Picosecond
REMPI	Resonance enhanced multiphoton ionization
SOA	Semiconductor optical amplifier
SS-OCT	Swept-source OCT
TDLAS	Tunable diode laser absorption spectroscopy
TD-OCT	Time domain OCT
TEA	Triethylamine
THz	Terahertz
Ti:sapphire	Titanium-doped sapphire
TOFMS	Time-of-flight mass spectrometry
VCSEL	Vertical cavity surface emitting laser
XRF	X-ray fluorescence analysis

1

Introduction

1.1 Use of Laser Diagnostics in Industrial Applications

Tighter environmental regulations in recent years have focused on harmful substances such as NO_x, CO, heavy metals, and particles, particularly from automotive and industrial sources. Anthropogenic carbon dioxide emissions are a major challenge and contribute to global warming and adverse effects on natural ecosystems. Demands for reversing environmental burdens grow steadily. It is thus becoming increasingly important to understand emission characteristics in order to minimize environmental disruption and to improve the efficiency of industrial machinery and plant processes. It is equally important to monitor controlling factors in order to improve the operation of industrial machinery and plants. In particular, detailed measurement techniques for the parameters of these factors, such as temperature and species concentration, are necessary to elucidate the overall nature of industrial systems. Plant conditions must be monitored in order to improve the operation of industrial systems, and improved on-line monitoring techniques of industrial controlling factors are required to enhance the controllability of overall system operations.

Currently, superfine particles are getting a lot of attention associated with nanotechnology research. Such small-size particles can easily enter into the human body and often settle in the lungs and deeper areas and are suspected to cause various diseases.[1.1] Nanosize particles in the atmosphere are mainly produced by the combustion of fossil fuels such as in engines, and artificially nanostructured supermolecules are intensively produced as highly functional materials. There is a high risk that highly functional materials are dispersed into the air from nanotechnology factories. Although nanosize particles make up only a small percentage of particle total mass, they account for a majority of the particle number (more than 90% of the particle number in diesel engine exhaust[1.2]), and their harmful influences on human health are being very seriously considered. It is rather difficult to estimate emissive conditions using conventional methods, and new evaluation criteria are required for a better understanding of their characteristics. It remains difficult to monitor the compositions and organic substances in nanoparticles because of the limited amount of nanoparticle constituents as well as their complexity.[1.3]

There are numerous standard methods to detect system parameters such as temperature and species concentrations, and these consist of well-known "industrial standards." Although these standard methods are well established and easily accessible, they often have limited performance in terms of meeting industrial needs because of slow response, complicated preconcentration procedures, and so on. In contrast, laser diagnostics make it possible to monitor these parameters because of their fast response, high sensitivity, and noncontact features. This fast response is critical for the control of industrial systems, and the noncontact detection is necessary for the clarification of their phenomena.

Industrial machinery and plants include engines, gas turbines, thermal and chemical plants, and disposal facilities. Industrial processes that contain chemical reactions, especially combustion, have been a major target in association with the so-called CO_2 problem. In car engines, an increasing concern in environmental issues such as air pollution, global warming, and petroleum depletion has helped drive research—including improvement of combustion controls such as variable valve timing and direct injection, and development of new power trains such as fuel cell and hybrid vehicles—in order to achieve low emissions and fuel consumption while maintaining engine power. For example, detailed analysis of combustion under transient conditions such as cold start is necessary to ensure lower emissions and fuel consumption. In order to realize this aim, it is necessary to use fast-response measurement techniques capable of analyzing temperature and concentration fluctuations in each engine combustion cycle. These requirements hold true of many processes such as gas turbines and thermal power plants.

There are mainly two types of approach to apply laser diagnostics to industrial applications. One is the clarification of basic phenomena in industrial processes. An example of this approach is the two-dimensional detection of temperature and species concentration using laser-induced fluorescence (LIF; see Chapter 3). LIF has been applied to noncontact detection of minor species and radicals.[1.4],[1.5] These applications include industrial burners, gas turbines, engines, and plasma devices.[1.5]–[1.22] Figure 1.1 shows a typical diagram of these applications. Noncontact and *in situ* measurement characteristics of laser diagnostics become an important part in these applications. The second approach is the monitoring and advanced control of industrial systems. Because of the advancement of laser devices, they have been reduced in size; built-in devices employing laser diagnostics have gradually been taking on a more useful shape. Examples of this approach are applications of laser-induced breakdown spectroscopy (LIBS; see Chapter 4) and tunable diode laser absorption spectroscopy (TDLAS; see Chapter 6). LIBS is suitable for elemental composition measurements in gas, liquid, and solid materials and has been intensively applied to various kinds of industrial plants, including iron and steel manufacturing processes, thermal power plants, and disposal facilities.[1.23]–[1.36] TDLAS enables the monitoring of process controlling parameters such as O_2, CO, CO_2, and NH_3 concentrations.[1.37]–[1.52] Figure 1.2

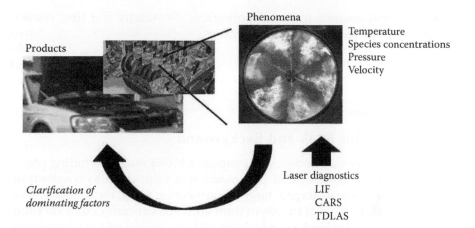

FIGURE 1.1
Application diagrams for clarification of the industrial process. Industrial systems are revised and modified through clarification of basic phenomena in industrial processes. Laser diagnostics is often suitable to figure out complicated phenomena because of its noncontact and *in situ* measurement features. (Courtesy of Professor Kidoguchi at the University of Tokushima.)

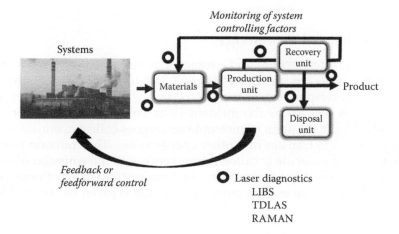

FIGURE 1.2
Application diagram for monitoring and advanced control of industrial processes. Measurement devices using laser diagnostics are installed in practical systems such as thermal power plants, and they are monitored and/or directly controlled in real time using readouts from the laser devices. The fast-response feature of laser diagnostics plays a major role in these applications.

shows a typical diagram of these applications. Sensitivity and time resolution are the main requirements for them.

1.2 Scope of This Book and Background

Applications of laser diagnostics encompass a broad range, including physics, engineering, medicine, and pharmacy. It is impossible to cover all of these areas, so this book largely targets engineering fields, especially practical applications. In view of trends to limit emissions of carbon dioxide from industrial systems, the book mainly deals with systems related to combustion and chemical reactions. These systems include engines, gas turbines, thermal and chemical plants, and disposal facilities. Accordingly, the laser diagnostics shown in this book are selected in terms of applicability to these systems. Brief explanations of these technologies and their backgrounds are shown in this chapter, with more specific details to follow in subsequent chapters. Laser-induced fluorescence is discussed in detail in Chapter 3, laser-induced breakdown spectroscopy in Chapter 4, spontaneous Raman spectroscopy and coherent anti-Stokes Raman spectroscopy (CARS) in Chapter 5, tunable diode laser absorption spectroscopy in Chapter 6, and time-of-flight mass spectrometry in Chapter 7. There are also growing biochemical application fields, and they are briefly introduced in Chapter 8.

1.2.1 Laser-Induced Fluorescence

An energy transfer process and geometric arrangement of laser-induced fluorescence (LIF) are shown in Figure 1.3.[1.4] In LIF the laser light with wavelength corresponding to the electronic energy difference of an atom or molecule is selected as the incident light. Following the absorption of the incident light, the atom or molecule undergoes collision, emission, and other processes to transfer into other energy states. The emission from an excited atom or molecule is called fluorescence, and the emission intensity gives information about temperature and concentration. The fluorescence intensity I_{LIF} from the excited atom or molecule is given by the following equation:

$$I_{LIF} = n_2 A_{21} h c \nu_{21} \Omega \varepsilon V / 4\pi \tag{1.1}$$

where n_2 is the number density of the excited atom or molecule, A_{21} the Einstein A coefficient, h the Planck's constant, c the velocity of light, ν_{21} the frequency of fluorescence light, Ω the solid angle of collection optics, ε the light collection

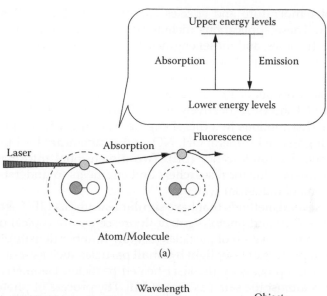

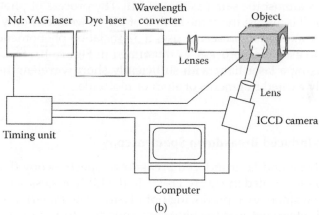

FIGURE 1.3

(a) Energy transfer process. Following the absorption of incident light, the molecule undergoes emission, collision, and other processes to transfer into other energy states. The emission is known as fluorescence and is used to determine concentration and temperature. (b) Typical geometric arrangement. The main components of a LIF system are a laser and a CCD camera. An ICCD camera is often used to pick out the fast LIF signal from other noise signals.

efficiency, and V the measurement volume. A typical setup of LIF consists of a laser (often tunable) and a charge-coupled device (CCD) camera. An image-intensified CCD camera (ICCD camera) is often used for the purposes of fast gate and signal intensification.

LIF is one of the most widely used laser diagnostics in analyses of physical phenomenon and has been applied mainly to minor species measurements.[1.4]–[1.21] LIF has been applied to two-dimensional (2-D) detection of temperature and species concentration due to its strong signal intensity. These applications include basic research burners, industrial burners, gas turbines, and diesel engines. Recently, three-dimensional (3-D) detection has also been shown to be capable of detecting several 2-D sections simultaneously. LIF has also been applied to flow and plasma diagnostics. Typical results of LIF are shown in Figure 1.4.[1.16] From the upper block, Figure 1.4 shows, respectively, direct flame images, OH, NO, soot, and temperature distributions in a diesel engine. The direct flame images were taken by a high-speed camera; OH, NO, and temperature by LIF; and soot by laser-induced incandescence (LII). Two-dimensional information often becomes a valuable clue for researchers and engineers to understand inherent phenomena of industrial systems.

There are several methods that have similar features as LIF. Laser-induced incandescence (LII) and photofragment fluorescence are typical ones. LII is used to obtain information of particles (usually number density).[1.5] LII uses an absorption process of laser light by small particles such as soot and a subsequent emission process by the laser-heated particles. Geometric arrangement of LII is almost the same as that of LIF. The process of photofragment fluorescence[1.5] is almost the same as that of LIF as well. The only difference is that photofragment fluorescence uses a dissociation phenomenon after (or instead of) an excitation process, as illustrated in Figure 1.5. Dissociation of molecules using a laser light with sufficiently short wavelengths produces electronically excited fragments of atom or molecule.

1.2.2 Laser-Induced Breakdown Spectroscopy

The principle behind laser-induced breakdown spectroscopy (LIBS) and its schematic are illustrated in Figure 1.6.[1.23] In the LIBS process, a laser beam is focused into a small area, producing hot plasma. The material contained in the plasma is atomized, and the light corresponding to a unique wavelength of each element is emitted from excited atoms in the plasma. The emission intensity from the atomized species provides information on elemental composition. The LIBS signal intensity I_{LIBS} from the atomized species has a similar formula to that of LIF and is related to the number density of the excited atom N_2 by the relation

$$I_{LIBS} = n_2 A_{21} hc\nu_{21} \Omega \varepsilon V / 4\pi \tag{1.2}$$

In LIBS n_2 is induced by a breakdown process instead of an absorption process in LIF. A typical setup of LIBS consists of a laser and a spectrometer.

LIBS is suitable for composition measurement in gas, liquid, and solid materials because of its strong signal intensity and simplicity of the apparatus.[1.23]–[1.36] LIBS applications have also been discussed intensively as process

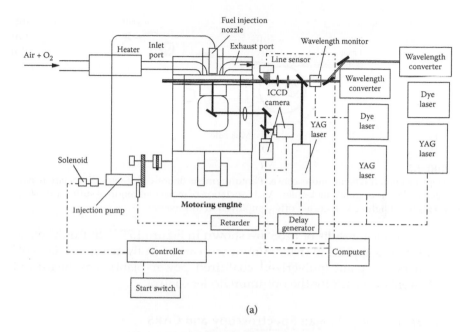

(a)

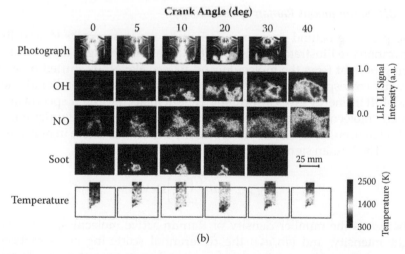

(b)

FIGURE 1.4

LIF and LII applications to engine combustion. (a) Experimental setup. Two dye laser systems were used to measure OH, NO, and temperature. Temperature was measured by two-line LIF of NO. The soot was measured using LII with 1064 YAG laser output. (b) LIF and LII measurement results. OH is present outside the region where the flame luminescence is observed; the reaction process is still taking place at the timing of 40° after top dead center where the flame luminescence is no longer observed. The NO distribution is located slightly outside the flame luminescence, in almost the same region as that of OH and high temperature. This fact corresponds to the formation process of NO hypothesized from an extended Zeldovich mechanism. Soot formation occurs in the fuel-rich region in the flame center and shows a trend similar to that of flame luminescence. (Reprinted from [1.16] with permission from IOP Publishing.)

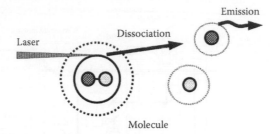

FIGURE 1.5
Concept of photofragment fluorescence. Photofragment fluorescence uses dissociation phenomena to detect "original" molecules. As dissociation of molecules requires sufficiently short wavelengths of light, UV lasers are often employed in this method.

monitors. Typical results of LIBS are shown in Figure 1.7.[1.32] In this application, LIBS was applied to detect unburned carbon in fly ash in a 1000 MW thermal power plant (pulverized coal-fired power plant). The measured results were also used for the optimum boiler control.

1.2.3 Spontaneous Raman Spectroscopy and CARS

1.2.3.1 Spontaneous Raman Spectroscopy

The principle behind spontaneous Raman spectroscopy[1.4],[1.5] and its configuration are illustrated in Figure 1.8. Raman scattering is the inelastic scattering of light from the molecule, and its wavelength is shifted by energies usually corresponding to molecular vibrational and/or rotational energies. The shift in energy gives information about molecules; their spectra are often called "molecular fingerprints." The wavelength shifts are unique for individual molecules, and multiple species detection is possible in many applications. The Raman signal intensity I_{Raman} is given by

$$I_{Raman} = nI_0 \left(\frac{d\sigma}{d\Omega} \right) \Omega \varepsilon V/4\pi \qquad (1.3)$$

where n is the number density of Raman-active molecules, I_0 the incident light intensity, and $(d\sigma/d\Omega)$ the differential scattering cross section. The interpretation of its signal and experimental setup are rather simple, and this makes spontaneous Raman spectroscopy an attractive method for industrial applications. On the other hand, spontaneous Raman spectroscopy is very weak and this limits its applicability. Fluorescence especially interferes often, making this method ineffective in many applications.

1.2.3.2 CARS

CARS is the acronym for "coherent anti-Stokes Raman spectroscopy," and it is one of the nonlinear optical processes using Raman effects.[1.4],[1.5],[1.37] The principle

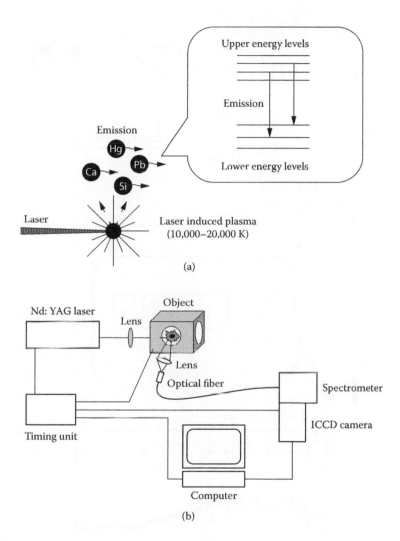

(a)

(b)

FIGURE 1.6
Principle and schematic of LIBS. (a) Plasma induced by a laser beam instantaneously reaches the 10,000–20,000 K temperature regions. The plasma first emits strong continuous noise, with atomic emissions appearing after a specific time delay. The necessary time delay is dependent on the upper energy of the measured element and plasma conditions. (b) Typical geometric arrangement. Main components of a LIBS system are a laser, a spectrometer, and a CCD camera. A LIBS signal is highly dependent on the plasma temperature, which means it depends on the delay time from the laser input. An ICCD camera is mostly used to select the preferable delay time for a measurement element.

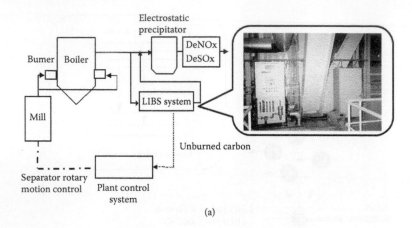

(a)

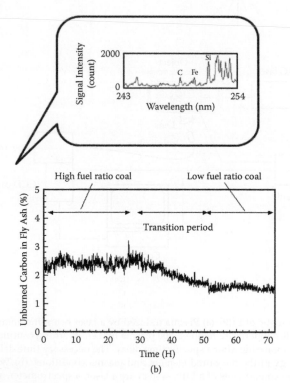

(b)

FIGURE 1.7

LIBS application to thermal plant control. (a) Thermal plant control scheme. Measured unburned carbon values were processed and used to control the mill. The mill rotary separator can control the particle size of the pulverized coal. The smaller the particle size, the lower the unburned carbon. (b) Measurement results of unburned carbon in fly ash. Unburned carbon in fly ash can be calculated from the major species concentrations in fly ash, which are Si, Al, Fe, Ca, and C. The transition from a high to a low fuel-ratio coal caused the change of unburned carbon values. (Reprinted from [1.32] with permission from Optical Society of America.)

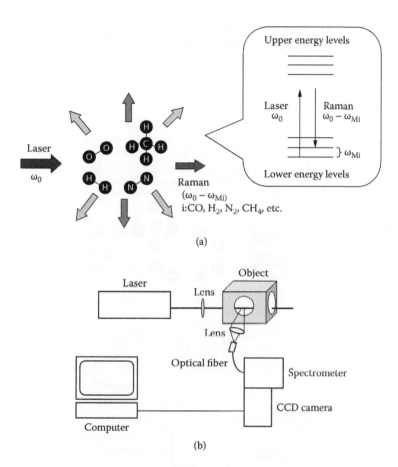

(a)

(b)

FIGURE 1.8
Principle and configuration of Raman spectroscopy. (a) Energy transfer process. The wavelength of Raman scattering light is shifted from that of the incident light by energies corresponding to molecular vibrational and/or rotational energies. The wavelength shifts are unique for individual molecules, and multiple species detection is possible in many applications. (b) Typical geometric arrangement. Main components of a Raman spectroscopy system are a laser, a spectrometer, and a CCD camera. The Raman scattering signal is very weak, and sensitive detectors are necessary to get sufficient signals.

behind CARS and its geometric arrangement are illustrated in Figure 1.9. CARS typically uses a pump beam ω_1 and a probe beam ω_2, and the blue-shifted (anti-Stokes) signal $\omega_3 = 2\omega_1 - \omega_2$ is generated by the nonlinear optical process. Different from other laser diagnostics, a CARS process forms a beamlike CARS signal, and this makes CARS unaffected by noise signals like fluorescence. Both spontaneous Raman spectroscopy and CARS can be used for temperature and species concentration measurements. The CARS signal intensity (I_{CARS}) is

$$I_{CARS}(\omega_3 = 2\omega_1 - \omega_2) = \varepsilon K \, |\chi_{CARS}|^2 \, I_1^2(\omega_1) I_2(\omega_2) \tag{1.4}$$

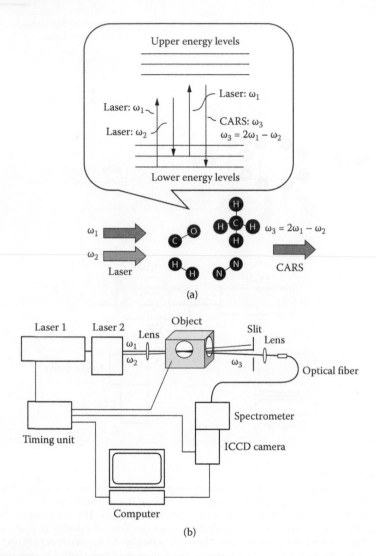

(a)

(b)

FIGURE 1.9

Principle and its configuration of CARS. (a) Energy transfer process. CARS is one of the non-linear optical processes. Using a pump beam ω_1 and a probe beam ω_2, the CARS signal with $\omega_3 = 2\omega_1 - \omega_2$ is generated by Raman effects. (b) BOXCARS geometric configuration. In CARS, two laser systems are usually employed for pump and probe beams. They are crossed with each other and a beamlike CARS signal is generated in the phase-matching condition.

where K is the absolute magnitude of CARS susceptibility, χ_{CARS} the CARS susceptibility, I_1 the pump beam intensity, and I_2 the probe beam intensity. As χ_{CARS} is proportional to the number density of Raman-active molecules n, the CARS signal intensity depends on n^2. Typical results of CARS are also shown in Figure 1.10.[1.38] Temperature probability density distribution was measured at the exhaust nozzle of a swirl-stabilized combustor.

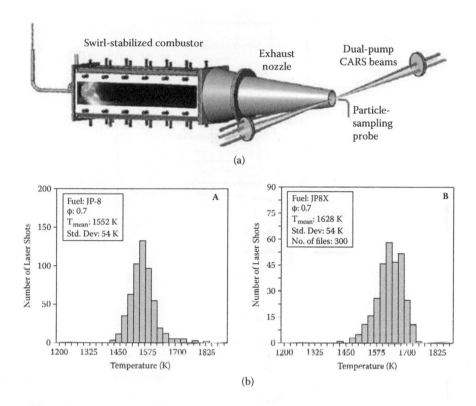

(a)

(b)

FIGURE 1.10
Typical results of CARS. (a) Schematic of swirl-stabilized combustor measurement. Pump and probe beams were focused in the exhaust gas of a swirl-stabilized combustor. Temperature and CO_2 concentration were measured for studies of jet fuel variants and particulate-mitigating additives. (b) Measurement results of temperature distribution. Probability density functions of the single-shot temperature measurements were measured for different equivalence ratios. The standard deviations of these temperature measurements were within 3–4% of the mean values. (Reprinted from [1.38] with permission from Elsevier.)

1.2.4 Tunable Diode Laser Absorption Spectroscopy

An energy transfer process and geometric arrangement of tunable diode laser absorption spectroscopy (TDLAS) are shown in Figure 1.11. When light permeates an absorption medium, the strength of the permeated light is related to the absorber concentration according to Lambert Beer's law. TDLAS uses this basic law to measure temperature and species concentration. It is also used to measure pressure and velocity. The number density of the measured species N is related to the amount of light absorbed I_ω, as in the following formula:

$$I_\omega/I_0 = \exp(-\kappa n \ell) \tag{1.5}$$

Here, I_0 is the incident light intensity, κ the absorption coefficient, and ℓ the pass length. The main feature of the TDLAS technique is its high sensitivity

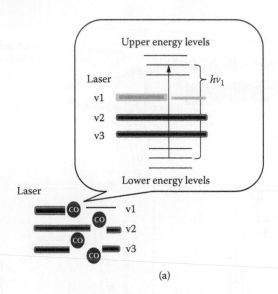

(a)

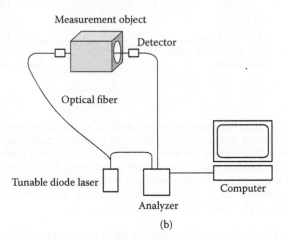

(b)

FIGURE 1.11
(a) Energy transfer process of TDLAS. When light permeates an absorption medium, the strength of the permeated light is related to absorber concentration according to Lambert Beer's law. Atoms and molecules have their own spectral pattern. Because of this feature TDLAS has excellent selectivity and sensitivity. (b) Typical geometric arrangement. The system of TDLAS is simple and its main components are a diode laser and a detector. Modulation methods or balanced receivers are often used to get small absorption signals.

[along the order of parts per billion (ppb)] and fast response (along the order of kHz). Extensive research[1.39]–[1.54] has been performed utilizing diode laser absorption spectroscopy for system monitoring and its control.

The high sensitivity and fast response of TDLAS enable the monitoring of system-controlling parameters such as O_2,[1.44] CO,[1.45] H_2O,[1.40],[1.44] CO_2,[1.45]

NH_3,[1.46] and NO.[1.47] As for combustion fields, its applications consist of not only small-scale test burners but also large-scale commercial combustors. Measurements with fast response and high sensitivity clear up the phenomena that cannot be detected by conventional measurement methods. Typical results of TDLAS are shown in Figure 1.12.[1.50] CO and O_2 were measured in a 300 ton/day commercial incinerator furnace for an advanced combustion control.

1.2.5 Time-of-Flight Mass Spectrometry

The principle behind the time-of-flight mass spectrometry (TOFMS) and its geometric arrangement are illustrated in Figure 1.13.[1.55] The measurement sample is introduced into the vacuum chamber and ionized by a laser ionization process, and electric field potential V_p is simultaneously applied for acceleration of generated ions. The accelerated ions enter a region with no potential difference (the drift region) and undergo uniform motion. At this moment, due to the law of energy conservation, the ions' electric field potential is equivalent to the kinetic energy, and a relationship is obtained such that the atomic or molecular mass of the sample is proportional to the square of the time of flight. Thus, the atomic or molecular weight can be determined by measuring the time of flight required for the ion to reach the detector. Due to the law of energy conservation, the ions' electric field potential is equivalent to their kinetic energy such that the following formula is established:

$$zV_p = \frac{1}{2}m\left(\frac{\ell}{t}\right)^2 \tag{1.6}$$

where z is the ion value, ℓ is the length of the drift region, m is the ion mass, and t is the time of flight of the ion traversing the drift region. In an application of laser diagnostics, a laser ionization process such as resonance-enhanced two-photon ionization is employed to reduce fragmentation.

The laser ionization TOFMS is capable of reaching the parts-per-trillion (ppt; 10^{-12}) level detection limit, and it has been demonstrated to be suitably sensitive for various hydrocarbons, chlorinated hydrocarbons such as dioxins[1.56],[1.57] and PCBs,[1.58],[1.59] and other substances. Typical results of TOFMS are shown in Figure 1.14.[1.60] The method was applied to on-line measurements at a waste incineration pilot plant. Many different polycyclic aromatic hydrocarbons (PAHs) were detected selectively from the complex flue gas matrix. The achieved detection limit for naphthalene is in the 10 ppt by volume.

1.2.6 Other Laser Diagnostics

There are other types of measurements in which laser diagnostics have been utilized in industrial applications. These include velocimetry, particle sizing, optical fiber sensing, and so on. They are mostly marketed by several equipment makers, and they are briefly reviewed in Chapter 8. There are also

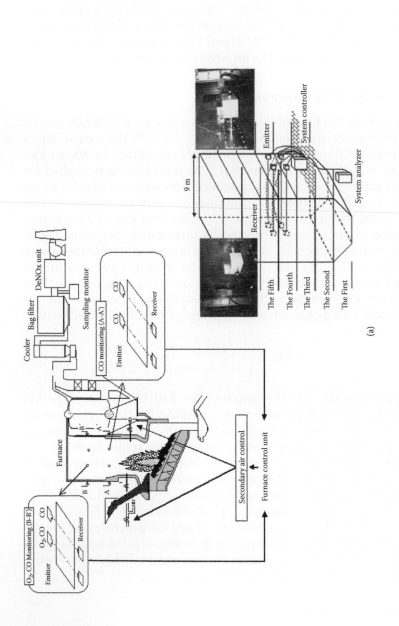

(a)

FIGURE 1.12

(a) Disposal plant control scheme in TDLAS. Measurement of CO concentration was conducted in the upper area of the secondary air, with O_2 measurement conducted from a point located further up. In secondary air allocation control, the allocation of the volume of secondary air is shifted based on the CO concentration ratios. (b) Comparison between TDLAS and conventional measurement results. The in-furnace laser measurements were capable of detecting the O_2 concentration fluctuations 2–3 minutes faster than the existing conventional monitor. This 2–3-minute period is extremely valuable for the control of combustion, because CO peaks arise in only 1–2 minutes. (Reprinted from [1.50] with permission from Elsevier.)

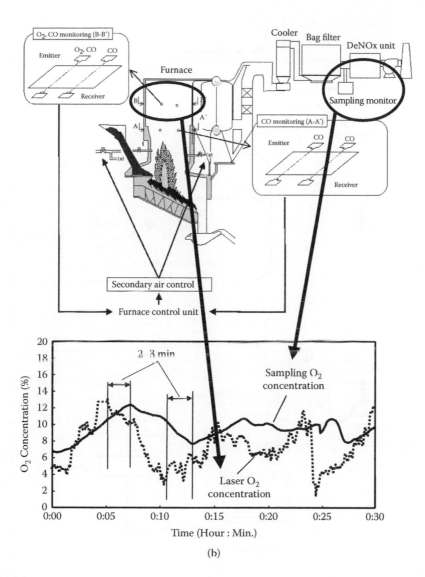

FIGURE 1.12
(Continued)

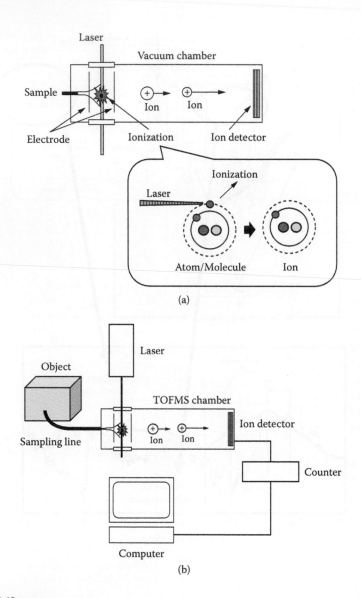

FIGURE 1.13
(a) Concept of TOFMS energy transfer process. Unlike other laser diagnostics, a laser ionization TOFMS method samples the gas to a vacuum chamber. In the TOFMS process, ions can be counted by the ion detector; therefore, super-high sensitivity can be achieved using this method. (b) Usual geometric arrangement. The main components are a laser and a TOFMS chamber. Supersonic jet cooling and ion trap methods are often used to enhance the sensitivity.

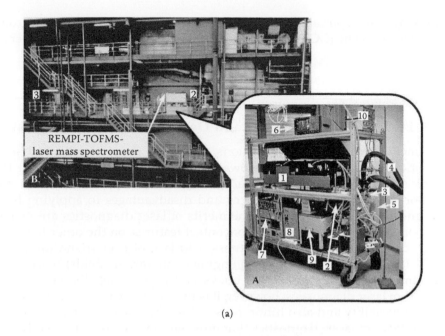

(a)

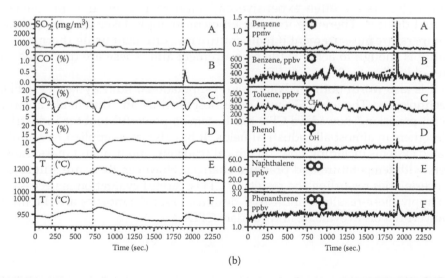

(b)

FIGURE 1.14

TOFMS application to disposal plant monitoring. (a) Pictures of a TOFMS system and its layout in a disposal plant. The components of the system were installed inside a box to place the system into a disposal plant. The exhaust gas was sampled and measured on site to clear the emission characteristics of a disposal plant. (b) PAH measurement results. TOFMS measurements were performed with the conventional on-line measurement techniques. Several PAHs were measured simultaneously. Compared with the conventional measurement methods (a–f), TOFMS measurement results (A–F) showed a sharp rise of individual PAH concentrations. (Reprinted from [1.60] with permission from Springer.)

growing laser application areas, such as terahertz technologies, optical computer tomography (OCT), and so on. They are also introduced in Chapter 8.

1.3 Evaluation of the Merits and Demerits on Laser Diagnostics

As mentioned in previous sections, there are many excellent characteristics of laser diagnostics. However, laser diagnostics are still at the early stages of development in terms of industrial applications, and it is important to know the advantages and disadvantages to applying laser diagnostics to practical fields. The merits of laser diagnostics are its fast response, high sensitivity, and noncontact features; on the other hand, its high cost, low liability or robustness, and lack of user friendliness (not only for spectroscopists but also engineers in various fields) work to its detriment. There are numerous books and papers that disseminate the merits of laser diagnostics; however, it is important to have a clear grasp of the applicability and also hindrances of laser diagnostics. One of the disadvantages of laser diagnostics that must be considered is its cost. Lasers and detectors are often expensive and vulnerable compared to conventional devices. Therefore it is not efficient to apply laser diagnostics for temperature measurement if temperature measurement using thermocoupling is enough to achieve the measurement aim. These are discussed in detail in Chapter 2.

It is also worth noting advances in computational fluid dynamics (CFD) technologies. Though CFD was once described as an "unreliable answer with precision," nowadays CFD has become a powerful tool for industrial designers in almost all industrial fields. It has also been applied to complicated systems including flows, chemical reactions, and heat and mass transfers with temperature and pressure changes. However, there are still some uncertainties regarding the results of CFDs; careful evaluation is necessary to apply these results to industrial designs. Laser diagnostics are the most powerful tools to evaluate and validate CFDs.

1.4 Laser Sources, Detectors, and Data Processors

Components of a measurement system play a key role in the success of industrial applications. Among them, laser devices have been the most important tool in laser diagnostics. Decades ago, lasers had an image of an "expensive and vulnerable scientific toy" for industrial engineers; however, compact and robust devices for both lasers and detectors are now available in a variety of forms. Advancement of semiconductor lasers is most notable for laser devices. Semiconductor lasers have made laser systems compact and robust, and they can also be used as a light source for various laser diagnostics.

For its part, what is unique about detectors is that various types of CCD image sensors have brought noteworthy improvement in sensitivity and detection time. Currently, two- and three-dimensional image detections of low-light signals (sometimes in the photon-counting region) are possible using highly sensitive CCDs. A brief summary of laser sources, detectors, and data processors is shown in Sections 1.41 and 1.42.

1.4.1 Laser Sources

Laser is the acronym for "light amplification by stimulated emission of radiation," wherein light broadly denotes electromagnetic radiation of any frequency. Therefore, the meaning of laser is originally "light amplification." The typical schematic of a laser is shown in Figure 1.15. Lasers inherently consist of an active medium, an excitation source, and cavity mirrors. The active medium usually contains the atoms or molecules that produce "stimulated emission" by the excitation source. The excitation sources, which are discharge, flash lamps, and often lasers, create a "population inversion" of the active medium, and a portion of laser radiation between cavity mirrors is delivered from one of the cavity mirrors (output mirror). "Laser" denotes a device that generates coherent light or the output of light itself, and the features of lasers are described in the following four main terms: monochromatic, coherent, directional, and short-pulse characteristics. There are several types of lasers that are useful in industrial applications. Lasers are also categorized according to their wavelength—for example, infrared, visible, ultraviolet, and x-ray lasers.

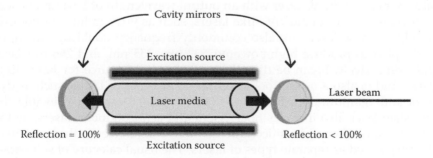

FIGURE 1.15
Typical configuration of laser. Lasers consist of an active medium, an excitation source, and cavity mirrors. The excitation sources create a "population inversion" of the active medium and lasing continues between cavity mirrors. A portion of laser radiation inside the cavity is delivered from the output mirror.

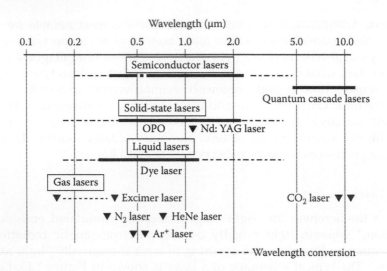

FIGURE 1.16
Laser types and output wavelength. Almost all wavelength regions from 200 nm to over 5000 nm are now covered by several types of laser devices.

The typical laser types that are used for industrial applications are shown in Figure 1.16.

1.4.1.1 Solid-State Lasers

In solid-state lasers, a laser medium consists of solid materials such as chromium-doped ruby and neodymium-doped yttrium aluminum garnet (Nd:YAG). The materials are usually made by doping a crystalline solid material with other appropriate elements. One of the most famous solid-state lasers is a Nd:YAG laser with an output wavelength of 1064 nm. (There are several families of Nd:YAG-like lasers, such as yttrium lithium fluoride [Nd:YLF].) These lasers are also commonly frequency doubled, tripled, or quadrupled to produce high-powered 532 nm, 355 nm, and 266 nm laser light, respectively. It can be directly used as a light source for laser diagnostics and is also a useful tool to pump other types of lasers, such as dye lasers and titanium-doped sapphire (Ti:sapphire) lasers. In principle the solid-state laser also includes fiber lasers and semiconductor lasers, as the active media—fiber and semiconductor—are solid; however, these are usually categorized as separate types of lasers. The usual category of solid-state laser usually excludes semiconductor lasers, which have their own name.

In industrial applications, solid-state lasers are one of the most useful light sources because of their "solid-medium" features. They are usually easier to use and more robust than liquid or gas lasers. Diode-pumped solid-state (DPSS) lasers are the most promising in many applications for their compact and robust characteristics. They also have longer lifetimes compared with flash-lamp pumped lasers. It is also possible to convert the laser wavelength using second-order nonlinear optical effects. The optical parametric oscillator (OSO) is one of these methods and can tune the laser wavelength such that the sum of the output light frequencies, which are called "signal" and "idler " frequencies, is equal to the input light frequency ω_0. This relation gives

$$\omega_0 = \omega_s + \omega_i \tag{1.7}$$

where ω_s is the signal frequency and ω_i the idler. The signal and idler frequencies can be tuned on the condition of Equation (1.7) by changing the angle of the nonlinear crystal.

1.4.1.2 Semiconductor Lasers

Advancement of semiconductor lasers is the most notable development in laser devices. Semiconductor lasers achieve the population inversion by injected electric current to a p–n junction.

As shown in Figure 1.17(a), holes (from the p-doped semiconductor) and electrons (from the n-doped semiconductor) are injected into the p n junction, and when they recombine the equivalent energy of the band gap is released in the form of photons. Amplified by the stimulated emission of radiation (photons) in cavity feedback, laser light can be obtained through the p–n junction area. There are different types of semiconductor lasers that emit at wavelengths from 375 nm to over 2000 nm. Quantum cascade lasers, which are categorized as semiconductor lasers and utilize semiconductor multiple quantum well heterostructures, emit in the mid- to far-infrared wavelengths region (over 3000 nm). Semiconductor lasers have been the light source of various laser diagnostics, especially tunable diode laser absorption spectroscopy, and have also been used to pump other lasers with high efficiency.

In tunable diode laser absorption spectroscopy, distributed feedback (DFB) lasers are most frequently employed, and their basic structure is shown in Figure 1.17(b). In DFB lasers, their resonators consist of a periodic structure or "grating," which acts as a distributed reflector in the wavelength range of laser emission. Because of this grating structure, typically one lasing mode is favored for light amplification, and a single-frequency laser output is delivered from the DFB structure. Changes of device temperature and input current cause the pitch of the grating to change, affecting the wavelength of the laser output. The typical wavelength tenability is about 1–2 nm for a device temperature control and 0.2–0.4 nm for input current. There are other types of tunable lasers utilizing a semiconductor laser.

An external cavity diode laser is one of these tunable lasers, and its output wavelength is stabilized by an external grating as feedback. The wavelength tuning range is usually wider than that of a DFB laser. Distributed Bragg

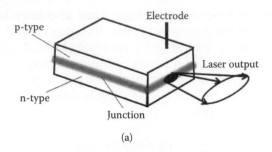

(a)

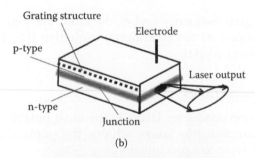

(b)

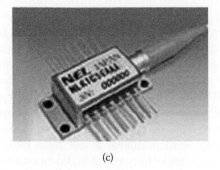

(c)

FIGURE 1.17
Principle of semiconductor lasers. (a) Structure of semiconductor laser. Laser light is emitted from the p–n junction area through the population inversion by injected electric current. (b) Structure of DFB laser. In DFB lasers, the "grating" structure is added inside the laser structure. Because of this structure, only one lasing mode is amplified along the p–n junction. (c) Picture of DFB laser. The size of the laser is approximately 30 × 33 × 8 mm. Courtesy of NEL Co.

reflector (DBR) lasers and vertical cavity surface emitting lasers (VCSELs) are also among the tunable diode lasers. The tuning range of this type of tunable lasers is rather limited and a multilaser system is necessary for wide wavelength tuning (that is, for multispecies detection). It is worth noting that tunable lasers with wavelength tuning over 100 nm and tuning speed in a kHz range are now possible by employing a semiconductor optical amplifier

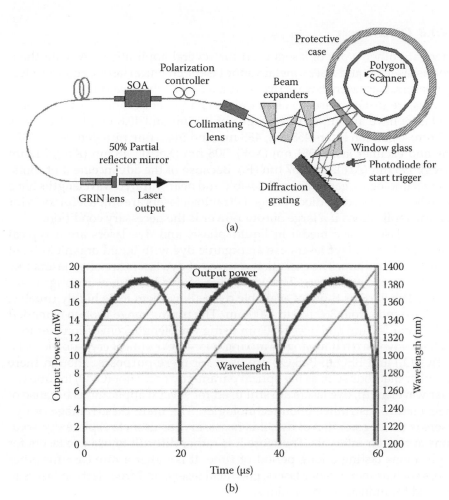

FIGURE 1.18
Structure of fast and wide wavelength scanning laser. The laser employed a polygon scanner and a semiconductor optical amplifier (SOA) to combine fast wavelength scanning with a wide wavelength range. The wavelength tuning range is 100 nm with 30 kHz scanning speed. (a) Structure of fast and wide wavelength scanning laser. (b) Common output power and scanning wavelength range. Courtesy of Santec Co.

(SOA) using an external grating. The schematic of this type of laser is shown in Figure 1.18. The wavelength tuning range of 100 nm with 30 kHz scanning speed can be achieved using a high-speed polygon scanner. Wavelength resolution is dependent on the wavelength scanning speed. These lasers have been used for an application of optical coherence tomography (OCT) and are also useful for tunable diode laser absorption spectroscopy in terms of multispecies detection.

1.4.1.3 Other Lasers

There are other types of lasers used for several applications. Among them are gas lasers, liquid lasers, metal vapor lasers, chemical lasers, and free electron lasers. Gas lasers have active media as a gas phase, and their excitation sources are usually gaseous electrical discharge. For example, carbon dioxide lasers have an output wavelength of 9.6 µm and 10.6 µm; He-Ne lasers 633 nm; argon-ion lasers 458 nm, 488 nm and 514.5 nm; nitrogen lasers 337.1 nm; and excimer lasers 351 nm (XeF), 308 nm (XeCl), 248 nm (KrF), 222nm (KrCl), 193 nm (ArF), and 157 nm (F_2). Because of the advancement of diode laser technologies, gas lasers with visible and near-infrared wavelengths have often been replaced by diode lasers. Ultraviolet lasers are still useful tools for several applications if a large output power is the necessary condition.

Liquid lasers have media in liquid phase, and dye lasers are a typical example of them. Dye lasers use an organic dye with liquid organic solvent as the gain medium. The excitation sources of dye lasers are often lasers like Nd:YAG or excimer lasers. The advantage of dye lasers is wavelength tenability. The various types of available dyes allow them to be highly tunable, which extends from 360 nm to 1000 nm. The tuning range is often extended both under 360 nm and over 1000 nm using wavelength conversion methods such as wavelength doubling or mixing. Nonlinear optical crystals like KTP ($KTiOPO_4$) and BBO (β-BaB_2O_4) are used for these purposes. Though there are other methods such as the optical parametric oscillator (OPO) to tune the laser wavelength, dye lasers are still used for several applications because of their mature instrumentation technologies. The main disadvantage of dye lasers is the degradation of dye; it is necessary for users to replace dye solutions at regular intervals. Therefore it is rather difficult to use dye lasers for applications lasting a long period of time. It is rather a rare case for other lasers such as metal vapor lasers, chemical lasers, and free electron lasers to be used for industrial applications.

1.4.2 Detectors

Decades ago, the photomultiplier with a monochromator was an essential tool for spectroscopy. Though it is still an important device for laser diagnostics, other devices such as CCDs have been intensively utilized in various fields. CCD image sensors have brought noteworthy improvement in sensitivity and detection time for several applications. Two- and three-dimensional image detections of low-light signals (sometimes in the photon-counting region) are now possible using highly sensitive CCDs. The typical detector types that are used for industrial applications are shown in Figure 1.19. Photomultiplier tubes, photodiodes, and CCD image sensors are discussed in more detail in this section.

In many cases detectors are used with wavelength-selective devices such as monochromators or filters. A monochromator is a device that transmits

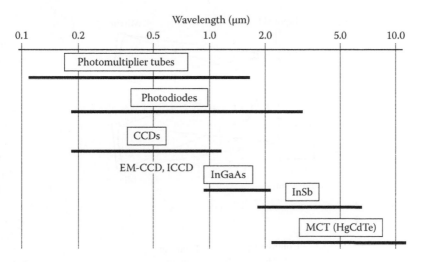

FIGURE 1.19
Detector types and detection wavelength.

a selected narrow band of wavelengths of light from a wider wavelength range. It often uses diffraction gratings for the wavelength separation. Its common configuration is shown in Figure 1.20(a). The Czerny–Turner configuration has been commonly employed for rather high-resolution applications, and its spectral resolution is determined by the width of the entrance and exit slits and the groove frequency given as number of grooves per millimeter. There are also other types of gratings and volume phase holographic (VPH) transmission gratings, which have notable characteristics. As shown in Figure 1.20(b), their structure usually includes a photosensitive gelatin between two substrates such as fused silica plates, and their index of refraction varies periodically throughout the volume of the grating. Their efficiency can exceed over 80%, and they are much more robust than conventional gratings because of their unique structure.

1.4.2.1 Photomultiplier Tubes

The operation of a photomultiplier tube (PMT) is shown in Figure 1.21(a). A typical PMT consists of a photocathode, focusing electrodes, an electron multiplier, and an anode sealed in a vacuum tube. Light introduced into the photocathode causes the emission of photoelectrons from it. This process occurs if the photon energy $h\nu$ exceeds the work function of the photocathode material, so there are several types of photomultiplier tubes according to the applicable wavelength. These photoelectrons are focused on the electron multiplier where the electrons are multiplied in a secondary emission process. This secondary emission process is repeated in the electron multiplier, creating 1 to 10 million times or more photoelectrons.

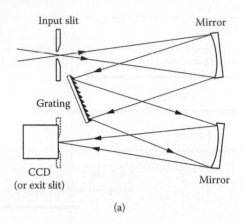

(a)

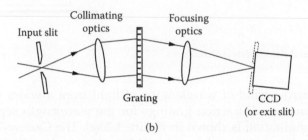

(b)

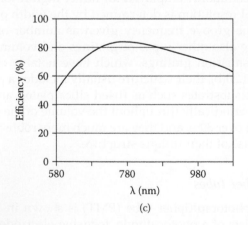

(c)

FIGURE 1.20

Common configuration for grating monochrometers. The efficiency of VPH gratings can exceed 80%, and they are much more robust than conventional gratings because of their unique structure. (a) Configuration of Czerny–Turner monochromator. (b) Common configuration of VPH grating spectrograph. (c) Typical efficiency curve of VPH grating. (Courtesy of P & P Optica Inc.)

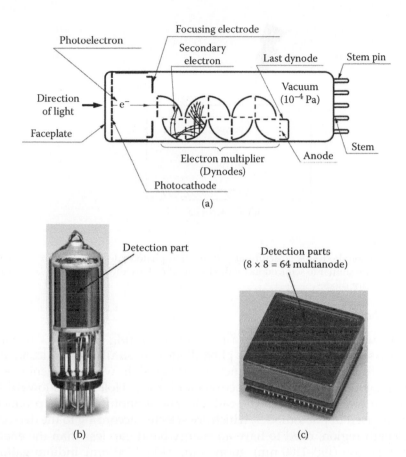

FIGURE 1.21
Principle of photomultiplier tubes (PMTs). Photomultiplier tubes multiply a photoelectron 1 to 10 million times or more inside a vacuum tube. Therefore they have features of high sensitivity and low noise. There are now several types of PMTs, such as flat panel and metal package PMTs. (a) Basic structure of photomultiplier tube. (b) Photo of normal side on photomultiplier tube. (c) Photo of flat panel photomultiplier tube. (Courtesy of Hamamatsu Photonics K.K.)

The detectable wavelength range of PMTs covers the entire range of 115–1700 nm depending on the photocathode materials, which include Cs-I (115–200 nm), Cs-Te (160–320 nm), Bialkali: Sb-K-Cs (160–650 nm), Multialkali: Sb-Na-K-Cs (160–900 nm), InGaAs (185–1010 nm), and InP/InGaAs (300–1700 nm). The PMT is highly sensitive and a photon counting method can be used to detect very weak optical signals. Its time resolution is also excellent, and it covers the ps time resolution. As the name "photomultiplier tube" implies, PMTs normally use the "tube" to seal their component in vacuum. However, there are now several types of PMTs, including flat panel PMTs and metal package PMTs.

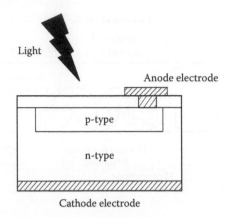

FIGURE 1.22
Principle of photodiodes. A structure of p–n junction photodiodes is shown in this figure. Photodiodes are similar to semiconductor diodes and they have the reverse process compared to semiconductor diodes.

1.4.2.2 Photodiodes

Photodiodes are photodetectors capable of converting light into current or voltage. As shown in Figure 1.22, photodiodes are similar to semiconductor diodes with a PIN junction or junctions. When light with sufficient energy enters photodiodes, it creates electrons and holes. Holes move toward the anode and electrons toward the cathode, and a photocurrent is produced. Materials used in photodiodes, which are selected according to the detection wavelength region, need to have an energy band gap less than the energy of light. Silicon (190–1100 nm), germanium (400–1700 nm), indium gallium arsenide (800–2600 nm), and lead sulfide (1000–3500 nm) are commonly used materials for photodiodes.

Photodiodes have many splendid features, which include low cost, low noise, excellent linearity, small size, low weight, longer lifetime, and high quantum efficiency (over 80 %). On the other hand, their sensitivity is rather low compared to that of PMTs, and they are usually used in applications with sufficient light intensities, such as tunable diode laser absorption spectroscopy.

1.4.2.3 CCD Image Sensors

Charge-coupled device (CCD) image sensors are a solid-state image sensor made of semiconductor devices. When converting the (light) image into electrical signals, a circuit called a "charge-coupled device" is used to read the charge generated from the light, as shown in Figure 1.20. CCD image sensors are sensitive and have less noise compared to other imaging devices. They are often used with cooling devices such as Peltier coolers to reduce so-called dark noise.

There are one- and two-dimensional image sensors in CCDs. In one-dimensional CCD sensors, also called linear CCD sensors, photodiodes are placed linearly with CCDs. One-dimensional sensors with a combination of monochromators are often employed to receive spectra, thus they are called spectrometers. Two-dimensional CCD sensors have photodiodes planarly placed with CCDs. A typical structure of two-dimensional CCD sensors is shown in Figure 1.23(a). They have been used for both spectrometers

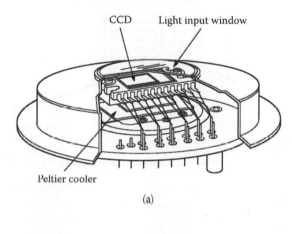

(a)

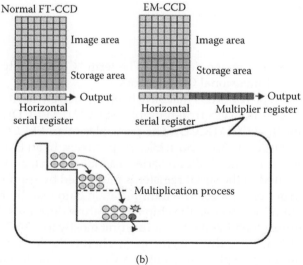

(b)

FIGURE 1.23
Structures of two-dimensional CCDs. Both EM-CCD and ICCD multiplied photoelectrons inside their structure to achieve high sensitivity. (a) Usual CCD structure. (b) Principle and structure of EM-CCD. (c) Principle and structure of ICCD. (Courtesy of Hamamatsu Photonics K.K.)

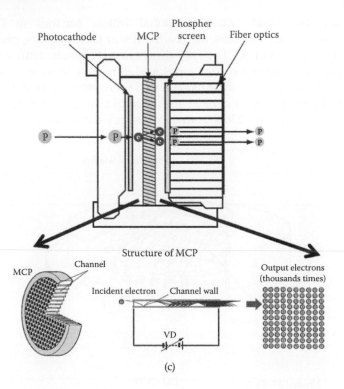

FIGURE 1.23
(Continued)

and two-dimensional image detectors. The term "CCD detector" is usually used for the two-dimensional CCD sensors.

There are two types of CCDs that have been intensively used for ultra-low-light detection. One is an electron-multiplying CCD (EM-CCD) and the other an image-intensified CCD (ICCD). The principles of an EM-CCD and ICCD are depicted in Figure 1.23(b) and 1.23(c), respectively. In an EM-CCD, a multiplication register is placed after a serial readout register, and the accumulated charge shifted to the serial register is multiplied by supplying a higher voltage to transfer electrodes. The multiplication gain usually reaches up to 1000 and the gain increases as the chip temperature decreases. Therefore its chip temperature is stabilized by a cooling unit mostly to −70°C. On the other hand, an ICCD includes three elements [a photocathode, a micro-channel plate (MCP), and a phosphor] in front of a CCD detector. It multiplies photoelectrons using the MCP, which can multiply photoelectrons by several thousand times and be operated by the fast-gate operation mode up to the ps region. Both EM-CCDs and ICCDs are highly sensitive and the ICCD is normally used for applications that require fast gating operations from the ps to μs regions. For applications with a longer accumulation timescale, the EM-CCD becomes the most preferable alternative.

Progress in computer design deserves special consideration for the advancement of laser diagnostics. Because of improvements in CPUs and memory capabilities, computing speed has become 2,300,000 times faster in 30 years, and this progress has offered great advantages in laser diagnostics especially for industrial applications. For almost all applications, data processing is necessary to convert measured data into desired forms such as temperature and concentrations. In case of a 1000 × 1000 CCD, 106 data points must be processed within a reasonable time. Data processing techniques such as fast Fourier transform (FFT) are often necessary to extract the maximum amount of measured results, and the computing speed becomes a critical requisite condition. This requirement is significant for the monitoring and advanced control applications in industrial systems.

References

[1.1]   K. Donaldson, X.Y. Li, and W. MacNee, "Ultrafine (nanometer) particle mediated lung injury," *Journal of Aerosol Science*, 29(5/6), 553–560, 1998.

[1.2]   D.B. Kittelson, "Engines and nanoparticles: A review," *Journal of Aerosol Science*, 29(5/6), 575–588, 1998.

[1.3]   H.J. Tobias and P.J. Ziemann, " Thermal desorption mass spectrometric analysis of organic aerosol formed from reactions of 1-tetradecene and O_3 in the presence of alcohols and carboxylic acids," *Environmental Science and Technology*, 34(11), 2105, 2000.

[1.4]   A.C. Eckbreth, *Laser Diagnostics for Combustion Temperature and Species*, Cambridge, Mass., Abacus Press, 1988.

[1.5]   K. Kohse-Hoinghaus and J.B. Jeffries, *Applied Combustion Diagnostics*, New York, Taylor and Francis, 2002.

[1.6]   D.R. Crosely, "Semiquantitative laser-induced fluorescence in flames," *Combustion and Flame*, 78, 153–167, 1989.

[1.7]   E.W. Rothe and P. Andresen, "Application of tunable excimer lasers to combustion diagnostics: A review," *Applied Optics*, 36(18), 3971–4033, 1997.

[1.8]   M. Knapp, A. Luczak, V. Beushausen, W. Hentschel, P. Manz, and P. Andresen, "Quantitative in-cylinder NO LIF measurements with a KrF excimer laser applied to a mass-production SI engine fueled with isooctane and regular gasoline," *SAE Technical Paper* 970824, 19–30, 1997.

[1.9]   K. Verbiezen, A.J. Donkerbroek, R.J.H. Klein-Douwel, A.P. van Vliet, P.J.M. Frijters, X.L.J. Seykens, R.S.G. Baert, W.L. Meerts, N.J. Dam, and J.J. ter Meulen, "Diesel combustion: In-cylinder NO concentrations in relation to injection timing," *Combustion and Flame*, 151(1/2), 333–346, 2007.

[1.10] M.G. Allen, K.R. McManus, D.M. Sonnenfroh, and P.H. Paul, "Planar laser-induced-fluorescence imaging measurements of OH and hydrocarbon fuel fragments in high-pressure spray-flame combustion," *Applied Optics*, 34(27), 6287–6300, 1995.

[1.11] K. Kohse-Hoinghaus, U. Meier, and B. Attal-Tretout, "Laser-induced fluorescence study of OH in flat flames of 1–10 bar compared with resonance CARS experiments," *Applied Optics*, 29(10), 1560–1569, 1990.

[1.12] H. Becker, A. Arnold, R. Suntz, P. Monkhouse, J. Wolfrum, R. Maly, and W. Pfister, "Investigation of flame structure and burning behavior in an IC engine simulator by 2D-LIF of hydroxyl radicals," *Applied Physics B*, 50(6), 473478, 1990.

[1.13] A. Arnold, B. Lange, T. Bouche, T. Heitzmann, G. Schiff, W. Ketterle, P. Monkhouse, and J. Wolfrum, "Absolute temperature fields in flames by 2D-LIF of hydroxyl using excimer lasers and CARS spectroscopy," *Berichte der Bunsen-Gesellschaft*, 96(10), 1388–1393, 1992.

[1.14] A.A. Rotunno, M. Winter, G.M. Dobbs, and L.A. Melton, "Direct calibration procedures for exciplex-based vapor/liquid visualization of fuel sprays," *Combustion Science and Technology*, 71(4–6), 247–261, 1990.

[1.16] Y. Deguchi, M. Noda, Y. Fukuda, Y. Ichinose, Y. Endo, M. Inada, Y. Abe, and S. Iwasaki, "Industrial applications of temperature and species concentration monitoring using laser diagnostics," *Measurement Science and Technology*, 13(10), R103–R115, 2002.

[1.17] S. Einecke, C. Schultz, and V. Sick, "Measurement of temperature, fuel concentration and equivalence ratio fields using tracer LIF in IC engine combustion," *Applied Physics B*, 71(5), 717–723, 2000.

[1.18] P. Pixner, R. Schiessl, A. Dreizler, and U. Maas, "Experimental determination of pdfs of OH radicals in IC engines using calibrated laser-induced fluorescence as a basis for modelling the end-phase of engine combustion," *Combustion Science and Technology*, 158, 485–509, 2000.

[1.19] S.R. Engel, P. Koch, A. Braeuer, and A. Leipertz, "Simultaneous laser-induced fluorescence and Raman imaging inside a hydrogen engine," *Applied Optics*, 48(35), 6643–6650, 2009.

[1.20] E.A. Brinkman, G.A. Raiche, M.S. Brown, and J.B. Jeffries, "Optical diagnostics for temperature measurement in a d.c. arcjet reactor used for diamond deposition," *Applied Physics B*, 64(6), 689–697, 1997.

[1.21] T. Kim, J.B. Ghandhi, "Investigation of light load HCCI combustion using formaldehyde planar laser-induced fluorescence," *Proceedings of the Combustion Institute*, 30(2), 2675–2682, 2005.

[1.22] S.R. Engel, P. Koch, A. Braeuer, and A. Leipertz ,"Simultaneous laser-induced fluorescence and Raman imaging inside a hydrogen engine," Applied Optics, 48(35), 6643–6650, 2009.

[1.23] A.W. Miziolek, V. Palleschi, and I. Schechter, *Laser Induced Breakdown Spectroscopy*, Cambridge, Cambridge University Press, 2008.

[1.24] L.W. Peng, W.L. Flower, K.R. Hencken, H.A. Johnsen, R.F. Renzi, and N.B. French, "A laser-based technique for continuously monitoring metal emissions from thermal waste treatment units," *Process Control and Quality*, 7(1), 39–49, 1995.

[1.25] S. Yalcin, D.R. Crosley, G.P. Smith, and G.W. Faris, "Spectroscopic characterization of laser-produced plasmas for in situ toxic metal monitoring," *Hazardous Waste and Hazardous Materials*, 13(1), 51–61, 1996.

[1.26] A.E. Pichahchy, D.A. Cremers, and M.J. Ferris, "Elemental analysis of metals under water using laser-induced breakdown spectroscopy," *Spectrochimica Acta Part B*, 52(1), 25–39, 1997.

[1.27] L. St-Onge, M. Sabsabi, and P. Cielo, "Analysis of solids using laser-induced plasma spectroscopy in double-pulse mode," *Spectrochimica Acta Part B*, 53(3), 407–415, 1998.

[1.28] R.E. Neuhauser, P. Panne, and R. Niessner, "Laser-induced plasma spectroscopy (LIPS): A versatile tool for monitoring heavy metal aerosols," *Analytica Chimica Acta*, 392(1), 47–54, 1999.

[1.29] C. Haisch, R. Niessner, O.I. Matveev, U. Panne, and N. Omenetto, "Element-specific determination of chlorine in gases by laser-induced-breakdown-spectroscopy (LIBS)," *Fresenius' Journal of Analytical Chemistry*, 356, 21–26, 1996.

[1.30] R.E. Neuhauser, P. Panne, R. Niessner, G.A. Petrucci, P. Cavalli, and N. Omenetto, "Online and in-situ detection of lead aerosols by plasma-spectroscopy and laser-excited atomic fluorescence spectroscopy," *Analytica Chimica Acta*, 346(1), 37–48, 1997.

[1.31] U. Panne, C. Haisch, M. Clara, and R. Niessner, "Analysis of glass and glass melts during the vitrification process of fly and bottom ashes by laser-induced plasma spectroscopy. Part I: Normalization and plasma diagnostics," *Spectrochimica Acta Part B*, 53(14), 1957–1968, 1998.

[1.32] M. Kurihara, K. Ikeda, Y. Izawa, Y. Deguchi, and H. Tarui, "Optimal boiler control through real-time monitoring of unburned carbon in fly ash by laser-induced breakdown spectroscopy," *Applied Optics*, 42(30), 6159–665, 2003.

[1.33] M. Gaft., E. Dvir, H. Modiano, and U. Schone, "Laser induced breakdown spectroscopy machine for online ash analyses in coal," *Spectrochimica Acta Part B*, 63, 1177, 2008.

[1.34] M. Pouzar, T. Cernohorsky, M. Prusova, P. Prokopcakov, and A. Krejcova, "LIBS analysis of crop plants," *Journal of Analytical Atomic Spectrometry*, 24(7), 953–957, 2009.

[1.35] M.A. Gondal, T. Hussain, Z.H. Yamani, and M.A. Baig, "On-line monitoring of remediation process of chromium polluted soil using LIBS," *Journal of Hazardous Materials*, 163(2–3), 1265–1271, 2009.

[1.36] O.T. Butler, W.R.L. Cairns, J.M. Cook, and C.M. Davidson , "Atomic spectrometry update. Environmental analysis," *Journal of Analytical Atomic Spectrometry*, 25(2), 103–141, 2010.

[1.37] S. Roy, J.R. Gord, and A.K. Patnaik, "Recent advances in coherent anti-Stokes Raman scattering spectroscopy: Fundamental developments and applications in reacting flows," *Progress in Energy and Combustion Science*, 36, 280–306, 2010.

[1.38] S. Roy, T.R. Meyer, R.P. Lucht, V.M. Belovich, E. Corporan, and J.R. Gord, "Temperature and CO_2 concentration measurements in the exhaust stream of a liquid-fueled combustor using dual-pump coherent anti-Stokes Raman scattering (CARS) spectroscopy," *Combustion and Flame*, 138(3), 273–306, 2004.

[1.39] M. Lackner, "Tunable diode laser absorption spectroscopy (TDLAS) in the process industries—A review," *Reviews in Chemical Engineering*, 23(2), 5–147, 2007.

[1.40] E.R. Furlong, D.S. Baer, and R.K. Hanson, "Real-time adaptive combustion control using diode-laser absorption sensors," *Symposium (International) on Combustion*, 27(1), 103–111, 1998.

[1.41] G. Winnewisser, T. Drascher, T. Giesen, I. Pak, F. Schmulling, and R. Schieder, "The tunable diode laser: A versatile spectroscopic tool," *Spectrochimica Acta Part A*, 55 (10), 2121–2142, 1999.

[1.42] S.W. Allendorf, D.K. Ottesen, D.R. Hardesty, D. Goldstein, C.W. Smith, and A.P. Malcolmson,"Laser-based sensor for real-time measurement of offgas composition and temperature in BOF steelmaking," *Iron and Steel Engineer*, 75(4), 31–35, 1998.

[1.43] P. Kohns, R. Stoermann, E. Budzynski, R.N. Walte, J. Knoop, and R. Kuester, "In-situ measurement of the water vapor concentration in industrial ovens by an user-friendly semiconductor laser system," *Proceedings of SPIE-The International Society for Optical Engineering*, 3098, 544–551, 1997.

[1.44] M.G. Allen, K.L. Carleton, S.J. Davis, W.J. Kessler, C.E. Otis, D.A. Palombo, and D.M. Sonnenfroh, "Ultrasensitive dual-beam absorption and gain spectroscopy: Applications for near-infrared and visible diode laser sensors," *Applied Optics*, 34(18), 3240–3249, 1995.

[1.45] D.B. Oh and D.C. Hovde, "Wavelength-modulation detection of acetylene with a near-infrared external-cavity diode laser," *Applied Optics*, 34(30), 7002–7005, 1995.

[1.46] D.M. Sonnenfroh and M.G. Allen, "Ultrasensitive, visible tunable diode laser detection of NO_2," *Applied Optics*, 35(21), 4053–4058, 1996.

[1.47] R.M. Mihalceal, M.E. Webber, D.S. Baer, R.K. Hanson, G.S. Feller, and W.B. Chapman, "Diode-laser absorption measurements of CO_2, H_2O, N_2O, and NH_3 near 2.0 µm," *Applied Physics B*, 67(3), 283–288, 1998.

[1.48] S.J. Carey, H. McCann, F.P. Hindle, K.B. Ozanyan, D.E. Winterbone, and E. Clough, "Chemical species tomography by near infra-red absorption," *Chemical Engineering Journal*, 77(1–2), 111–118, 2000.

[1.49] V. Ebert, J. Fitzer, I. Gerstenberg, K.U. Pleban, H. Pitz, J. Wolfrum, M. Jochem, and J. Martin, "Simultaneous laser-based in situ detection of oxygen and water in a waste incinerator for active combustion control purposes," *Symposium (International) on Combustion*, 27(1), 1301–1308, 1998.

[1.50] Y. Deguchi, M. Noda, M. Abe, and M. Abe, "Improvement of combustion control through real-time measurement of O_2 and CO concentrations in incinerators using diode laser absorption spectroscopy," *Proceedings of the Combustion Institute*, 29(1), 147–153, 2002.

[1.51] S. Barrass , Y. Gérard , R.J. Holdsworth, and P.A. Martin," Near-infrared tunable diode laser spectrometer for the remote sensing of vehicle emissions" *Spectrochimica Acta Part A*, 60(14), 3353–3360, 2004.

[1.52] M. Lewander, Z.G. Guan, L. Persson, A. Olsson, and S. Svanberg,"Food monitoring based on diode laser gas spectroscopy," *Applied Physics Part B*, 93(2–3), 619–625, 2008.

[1.53] M. Yamakage, K. Muta, Y. Deguchi, S. Fukada, T. Iwase, and T. Yoshida, "Development of direct and fast response gas measurement," *SAE Paper* 20081298, 51–59, 2008.

[1.54] C. Wang and P. Sahay, "Breath analysis using laser spectroscopic techniques: Breath biomarkers, spectral fingerprints, and detection limits," *Sensors*, 9, 8230–8262, 2009.

[1.55] D.M. Lubman, ed., *Lasers and Mass Spectrometry*, New York, Oxford University Press, 1990.

[1.56] R. Zimmermann, U. Boesl, C. Weickhardt, D. Lenoir, K.-W. Schramm, A. Kettrup and E.W. Schla, "Isomer-selective ionization of chlorinated aromatics with lasers for analytical time-of-flight mass spectrometry: First results of polychlorinated dibenzo-p-dioxins (PCDD), biphenyls (PCB) and benzenes (PCBz)," *Chemosphere*, 29(9), 1877–1888, 1994.

[1.57] H. Oser, R. Thanner, and H.-H. Grotheer, "Continuous monitoring of ultratrace products of incomplete combustion during incineration with a novel mobile JET-REMPI device," *Chemosphere*, 37(9), 2361–2374, 1998.

[1.58] Y. Deguchi , S. Dobashi, N. Fukuda, K. Shinoda, and M. Morita, " Real time PCB monitoring using time of flight mass spectrometry with picosecond laser ionization," *Environmental Science and Technology*, 37(20), 4737–4742, 2003.

[1.59] S. Dobashi, Y. Yamaguchi, Y. Izawa, Y. Deguchi, A. Wada, and M. Hara, "Laser mass spectrometry: Rapid analysis of polychlorinated biphenyls in exhaust gas of disposal plants," *Journal of Environment and Engineering*, 2(1), 25–34, 2007.

[1.60] R. Zimmermann, H.J. Heger, A. Kettrup, and U. Nikola, "Direct observation of the formation of aromatic pollutants in waste incineration flue gases by on-line REMPI-TOFMS laser mass spectrometry," *Fresenius' Journal of Analytical Chemistry*, 366(4), 368–374, 2000.

2

Application Codes of Laser
Diagnostics to Industrial Systems

2.1 Strength and Weakness of Laser Diagnostics

Laser diagnostics have attracted a great deal of attention in various industries because of the noncontact, fast response, and multidimensional features. There are numerous papers in which the excellent features of laser diagnostics have been demonstrated in practical conditions. However, in commercial plants and industrial systems such as turbines, boilers, engines, and so on, there are few measurement devices using laser diagnostics that have been used under actual operating conditions. The need for optical access (measurement windows) is also one of the negative factors for the practical applications of laser diagnostics. It is true that laser diagnostics have been demonstrated in a various fields to elucidate the physical phenomena in industrial systems, and many of them have played significant roles in the development of industrial systems. Although laser diagnostics are not required tools for machine designers to improve their systems, computational fluid dynamics (CFD) has become a powerful tool for designers in almost all industrial fields. It has also been applied to complicated systems including flows, chemical reactions, and heat and mass transfers. Laser diagnostics are still at an early stage of development in terms of industrial applications.

The gap between the potential of laser diagnostics and the actual applications has been mainly due to the costly, vulnerable, and difficult-to-use devices. Lasers were expensive instruments. Currently, advances in laser technologies, especially semiconductor lasers, are notable and have made lasers compact and robust. Lasers have been actually used in industries such as material processing and telecommunications. As applications of laser diagnostics to practical fields make further progress, it is important to know their advantages and disadvantages to ensure their maximum potential. Compared to conventional methods, laser diagnostics have:

- Advantages: Fast response, excellent spatial resolution, high sensitivity, and availability of *in situ* and noncontact measurements
- Disadvantages: High cost, low reliability, vulnerability, and difficult-to-use devices

These characteristics have been demonstrated in various applications, and laser diagnostics have the potential to improve industrial systems. However, we have to be sensitive in evaluating merits of laser diagnostics. Unlike lasers, thermocouples are affected by radiation and conduction phenomena. It is important to precisely evaluate these effects in academic applications. In cases where temperature fluctuations by radiation and conduction are not important to the products, laser diagnostics cannot compete with thermocouples. One of the most important aspects in industrial applications of laser diagnostics is the influence of measurement results to its products and systems.

There is another important factor to evaluate the merits of laser diagnostics against conventional techniques. Advances in conventional techniques have to be considered. Usually, it takes two or three years to develop laser measurement techniques, and the evaluation of laser diagnostics must be made on the grounds that conventional techniques also advance during this period. For example, there is an improvement of gas sampling methods; faster sampling techniques will be developed to achieve better response time in two or three years. The comparison of response time in concentration measurements have to be made by considering these advances.

It is also necessary to understand features of individual laser diagnostics to choose the right method in each application. Table 2.1 shows the characteristics of laser diagnostics considered in this book. Temperature can be measured using laser-induced fluorescence (LIF), tunable diode laser absorption spectroscopy (TDLAS), spontaneous Raman spectroscopy, and coherent anti-Stokes Raman spectroscopy (CARS). The most famous temperature measurement device is a thermocouple. In cases where the two-dimensional (2-D) temperature measurement is needed, LIF is the primary candidate for this application—but if a thermocouple is sufficient to measure temperature in an industrial process, a thermocouple is the best method for the application because of its lower cost and reliability. It is an easy and clear answer, but laser diagnostics have been used in the field where conventional measurement methods also work well. In such cases, laser diagnostics have big drawbacks in terms of cost and reliability.

There are mainly two types of approaches when applying laser diagnostics to industrial applications. They are usually called "seed-oriented" (measurement) and "need-oriented" (requirement) approaches. Figure 2.1 shows an example of these approaches. Though both are important to apply laser

TABLE 2.1

Characteristics of Laser Diagnostics

Method	Measurement Item	Measurement Dimension	System Cost	Reliability of Laser	Calibration Based on Theories	Multi-Species Detection	In Situ Measurement	Low Pressure	High Pressure	Detection Limit
LIF	Temperature concentration (velocity, pressure)	2-D	Medium-high	Poor	Difficult	Poor	Excellent	Excellent	Medium	ppm
LIBS	Elemental composition	Point	Medium	Medium	Difficult (calibration curve)	Excellent	Excellent	Medium	Medium	ppb-ppm
Spontaneous Raman Spectroscopy	Temperature concentration	1-D	Medium	Medium-Excellent	Easy	Excellent	Excellent	Poor	Excellent	%
CARS	Temperature concentration	Point	High	Poor	Medium	Poor	Excellent	Poor	Excellent	(ppm-)%
TDLAS	Temperature concentration (velocity)	Line of sight (2-D using CT)	Low	Excellent	Easy (self-calibrating)	Poor	Excellent	Excellent	Poor	ppb-ppm
LI-TOFMS	Concentration	Point	High	Poor	Easy (internal standardization)	Excellent	Based on sampling	Based on sampling	Based on sampling	ppt-ppb

Note: It is necessary to understand features of individual laser diagnostics to choose the right method in each application. Temperature can be measured using LIF, TDLAS, spontaneous Raman spectroscopy, and CARS, and the applied measurement method has to be chosen based on a need-oriented approach.

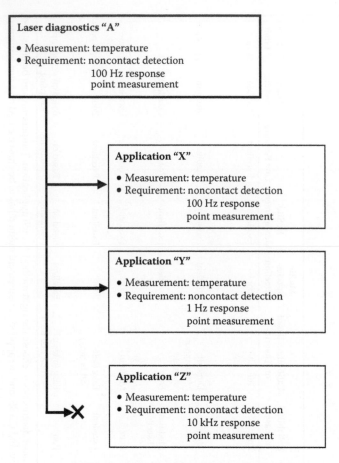

(a) "Seed-oriented" and "need-oriented" approach

FIGURE 2.1
Seed-oriented and need-oriented approaches. Though both are important in applying laser diagnostics to practical fields, need-oriented thinking is necessary for industrial applications. In a seed-oriented approach, a measurement method is already chosen and its applicability becomes the major issue. (a) Seed-oriented and need-oriented approach. (b) Need-oriented approach.

diagnostics to practical fields, need-oriented thinking is necessary for industrial applications. In a seed-oriented approach, a measurement method has already been chosen, and its applicability becomes the major issue. What is important in industry is not to apply a new method but to have a useful result. It is true that demonstrations of laser diagnostics in practical fields are important to show their ability and potential, even if laser diagnostics are not the best measurement technique from an industrial point of view. However, in the actual applications, laser diagnostics have to overcome the "Darwinian sea."

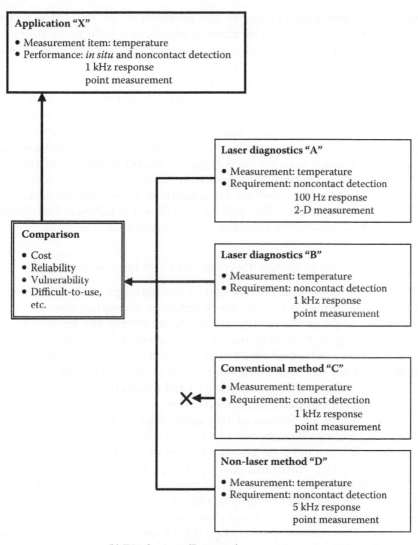

(b) "Need-oriented" approach

FIGURE 2.1
(Continued)

2.2 Industrial Applications of Laser Diagnostics

There are mainly two cases where the application of laser diagnostics to industrial applications is beneficial: to clarify basic phenomena in industrial processes in order to improve industrial products and systems, and in the monitoring and advanced control of industrial systems using laser diagnostics. Laser diagnostics has the potential to be applicable to both cases; however, requirements for those applications are different. In the former case, the important features are the top performances in measurement conditions: fast response, multidimensional, and noncontact characteristics. Ruggedness and reliability are not expected factors. The application ends in success by measuring the best results—even if these results can be detected at a rate of less than 20 percent of the total measurements. The majority of applications using laser diagnostics are categorized as these. However, the required characteristics of the latter applications—monitoring and advanced control of industrial systems—are ruggedness and reliability of the measurement systems. Though typical features of laser diagnostics such as fast response and noncontact abilities are important, these characteristics must be backed up by reliability. The application will end in failure if the measurement system does not work within specification in a short space of time.

2.2.1 Applications to System Development and Improvement

The identification of basic phenomena in industrial processes is an important and necessary step for the development of industrial systems. These basic studies improve the quality of industrial systems and prevent trouble from happening continuously. As for combustion systems such as engines and gas turbines, measured results have often been used to clarify the reaction mechanisms of combustion with the goal of reducing NO formation. This process is shown in Figure 1.1. Measured results have also been used as validation data of computer simulation codes.

The most important feature of this application is the top performances of laser diagnostics. Though robustness and reliability are also an important factor, they are not a necessary element in the development of industrial systems. After clarifying basic phenomena in industrial processes, countermeasures are taken using an appropriate technique other than laser diagnostics, because laser diagnostics is a measurement tool and does not directly improve the process. Laser diagnostics discussed in this book—LIF, laser-induced breakdown spectroscopy (LIBS), spontaneous Raman spectroscopy, CARS, TDLAS, and laser ionization time-of-flight mass spectrometry (LI-TOFMS)—can be used for these applications. Demands for laser diagnostics in measurement abilities tend to become higher, such as the requirements of a measurement dimension from a point measurement to 1-D, and then from 1-D to 2-D, and finally from 2-D to 3-D. These demands have been a motivation to improve performance of laser diagnostics. A typical example of this approach is the two-dimensional detection of temperature and species concentration using LIF (see Chapter 3). LIF has been applied to measure 2-D information of minor species and temperature. LIF has also been utilized to detect 3-D information.

The important thing in these applications is an evaluation of measurement characteristics by preliminary experiment and theory. Theoretical investigations are especially important because it is often difficult to do preliminary experiments under all measurement conditions. Laser diagnostics is based on this theory, and it is often possible to predict effects of conditions on the signal. LIF, TDLAS, spontaneous Raman spectroscopy, and CARS are the typical methods whose theories work well to predict temperature and pressure effects. Figure 2.2 shows H_2O absorption spectra in several temperature and pressure conditions. It is easy to understand the trend of absorption spectra in different temperature and pressure conditions using synthetic spectra; however, it is really difficult to get these trends by experiments. In TDLAS, laser wavelengths are often chosen by these theoretical calculations. It may be no exaggeration to say that these theoretical calculations are "maps to the measurement goal" and without these maps it will take a long time and unnecessary effort to achieve the goal. Figure 2.3 shows the outline of these procedures. The detailed procedures of each laser diagnostic are described in the following chapters.

2.2.2 Applications to Process and Environmental Monitoring and Control

The monitoring and advanced control of industrial systems using laser diagnostics have been the greatest challenges to its practical application. In these cases, ruggedness and reliability with a reasonable cost are the most important requirements, and they have also been a difficult challenge for laser diagnostics. Current advances in laser technologies are beginning to change this concept and make it possible to apply laser diagnostics to these fields. Lasers have been applied actively in many fields, especially telecommunications and laser processing. These lasers have been designed to have the ruggedness and reliability needed to meet industrial requirements and are now available in many applications.

Plant control using laser measurement devices is a typical example of this category. This process is shown in Figure 1.2. There are two types of trends in these applications. One is the continuous monitoring of important parameters in plants and industrial facilities. The other is measurement by a portable measurement device that can be used as needed at various sites. They are highly appealing to industry because laser diagnostics directly improve industrial processes. In these applications, all the components in a measurement system have to be reliable enough under the conditions of use. A typical example of this approach is the temperature and species concentration monitoring using TDLAS (see Chapter 6). The most vulnerable and necessary component is the laser, and the reason TDLAS is suitable for this application is apparent in Table 2.1. TDLAS often uses lasers that have been developed for telecommunications, which often have a usable lifetime of more than 10 years. The important thing in these applications is an examination of the entire set of components in the measurement system. In practical applications, it is not an easy task to change a laser unit every two weeks. There is no question that the measurement ability has to be checked before actually applying the method in the field.

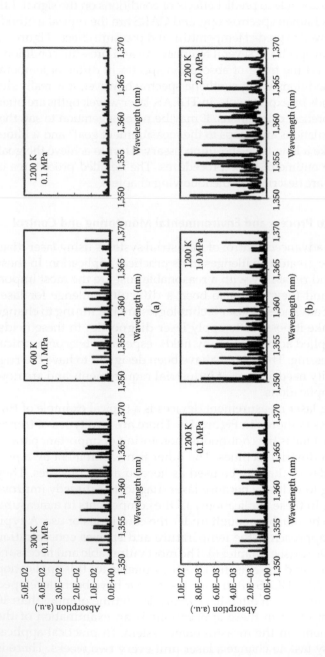

FIGURE 2.2

Theoretical H_2O absorption spectra in several temperature and pressure conditions. Theoretical investigations are especially important because it is often difficult to do preliminary experiments under all measurement conditions. Laser diagnostics is based on this theory, and it is often possible to predict effects of conditions to the signal. It is easy to understand the trend of absorption spectra in different temperature and pressure conditions using synthetic spectra; however, it is very difficult to get these trends by experiments. (a) Seed-oriented and need-oriented approach. (b) Need-oriented approach.

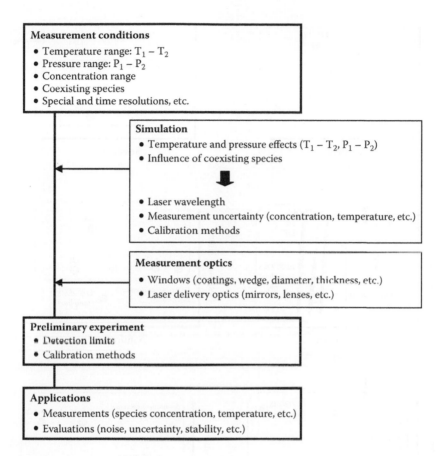

FIGURE 2.3
Outline of approach to industrial applications. Theoretical calculations are "maps" to a measurement goal, and they are necessary for many applications.

2.3 Application Guides to Industrial Systems

As discussed above, it is important to understand the feature of each laser diagnostic. Figure 2.4 shows the evaluation of each laser diagnostic in terms of factors that are important for industrial applications. LIF is an excellent method to measure temperature and species concentrations, and it can detect 2-D information using a charge-coupled device (CCD) camera. It is a good choice to use LIF with tunable lasers for the clarification of flow or reaction mechanism, although it is not recommended for use for continuous boiler control even if temperature information is valuable for boilers. Tunable lasers used in LIF are not reliable enough to be employed for long-term use; such kinds of plans almost always promise more than they deliver.

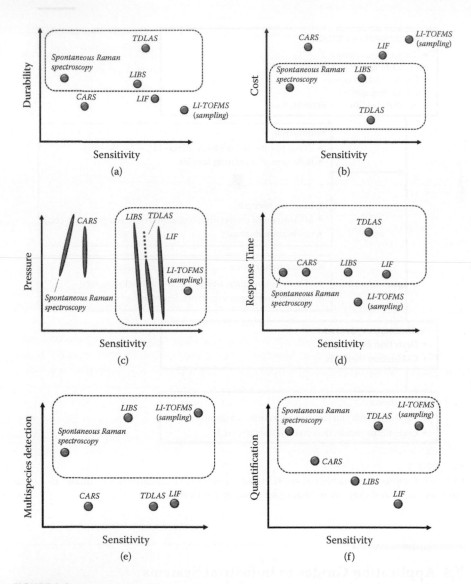

FIGURE 2.4

It is important to understand the features of each laser diagnostic. There are mainly two types of approaches to apply laser diagnostics to industrial applications, and requirements for them are different. (a) Durability—sensitivity. (b) Cost—sensitivity. (c) Pressure—sensitivity. (d) Response time—sensitivity. (e) Multispecies detection—sensitivity. (f) Quantification—sensitivity.

It is also important to be sensitive to advances in laser diagnostics. Decades ago CARS was a measurement technique that needed several lasers and complicated optical alignments. Now a single-beam CARS using a femtosecond (fs) laser is being developed, and the concept of CARS may change dramatically.

However, fs-lasers are still a complicated device, and further advances in laser technologies to use CARS for monitoring and control in industry are needed.

2.3.1 Application of Laser Diagnostics to Engine Systems

In car engines, an increasing concern in environmental issues such as air pollution, global warming, and petroleum depletion has helped drive research into various ways to employ laser diagnostics, especially in engine research. Engines are relatively small in size, and this is beneficial for many laser technologies, most of which are combustion analyses inside engine cylinders. In these applications there appear to be large temperature and pressure changes during the combustion cycles. Although there are some difficulties in applying laser diagnostics to these fields, applications have extended to both car engines (diesel and gasoline) and jet engines. Figure 2.5, Figure 2.6, and Figure 2.7 show application maps of laser diagnostics to engine systems. They are discussed in more detail in Sections 2.3.1.1, 2.3.1.2, and 2.3.1.3.

2.3.1.1 Air Intake Pipe

Figure 2.5 shows application diagrams of laser diagnostics in air intake pipes. There are few applications of laser diagnostics in this area compared to those in engine cylinders or exhaust pipes.

- TDLAS[2.1]

  TDLAS has been applied to the engine suction gas (intake air) measurement to clarify the exhaust gas recirculation (EGR) processes. The blowback of CO_2 can be clearly measured as a function of valve overlap time between intake and exhaust.

2.3.1.2 Engine Cylinder

Figure 2.6 shows application diagrams of laser diagnostics in engine cylinders. There have been numerous applications of laser diagnostics to elucidate the flow and combustion mechanisms in engines.

- LIF[2.2]–[2.7]

  There have been a numerous applications of LIF in engine combustions. One of the powerful methods of LIF is tracer-LIF in which molecules such as biacetyl, acetone, and formaldehyde are used as tracers. This method is often used to analyze unburned fuels or oxidizers. It is important to know the tracer characteristics before applying this method. Measurement results have been used to elucidate the jet fuel patterns and its mixing process during injection and the propagation of the flame during

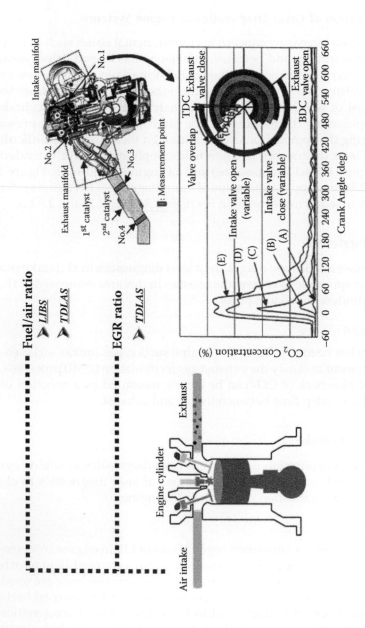

FIGURE 2.5
Application diagrams of laser diagnostics in air intake pipes. There are few applications of laser diagnostics in this area compared to those in engine cylinders or exhaust pipes. TDLAS has been applied to the engine suction gas (intake air) measurement to clarify the EGR processes.

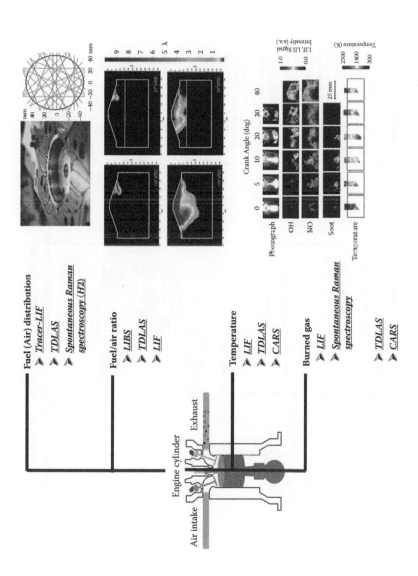

FIGURE 2.6

Application diagrams of laser diagnostics in engine cylinders. There have been a numerous applications of laser diagnostics to elucidate the flow and combustion mechanisms in engines. LIF, LIBS, spontaneous Raman spectroscopy, CARS, and TDLAS have been extensively applied to combustion analyses in engine cylinders.

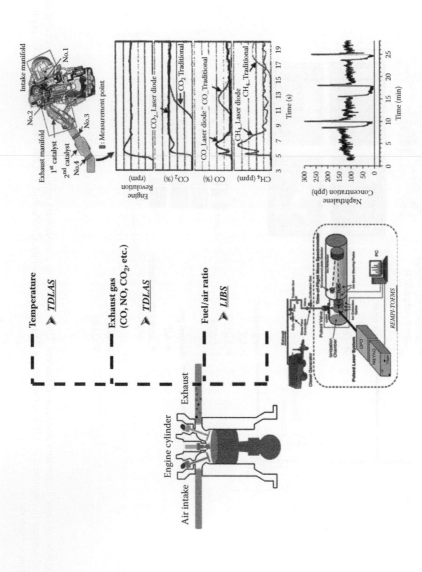

FIGURE 2.7

Application diagrams of laser diagnostics in exhaust pipes. Exhaust gas management is important for engines, and it is necessary to measure exhaust gas compositions with sufficient time resolution to analyze transient engine phenomena. LIF, LIBS, and LI-TOFMS have been used for this purpose.

combustion. These results can also be used for a direct optimization of mixture formation processes and the validation of CFD models.

LIF has been demonstrated to detect atoms or molecules that exist naturally in engine combustions. It is the most common application of LIF in engines. The species detected include OH, NO, and O_2. In addition to evaluating these molecules, soot is often measured using LII because soot is an important product of engine combustion. The main purposes of these applications are the reduction of NO_x and soot.

- LIBS[2.8],[2.9]

  LIBS has been used to measure fuel-air ratio in combustion. If a composition of fuel does not change, the fuel-air ratio can be inferred from elemental analyses of unburned or burned gases. The elemental analysis of unburned gases is mainly employed to evaluate fuel-air ratio of fuel-air mixtures.

- Spontaneous Raman spectroscopy and CARS[2.10],[2.11]

  Spontaneous Raman spectroscopy and CARS have been used for in-cylinder measurements. Especially CARS has been the tool for temperature measurement. There is a special advantage for spontaneous Raman spectroscopy and CARS compared to LIF and TDLAS, which is the detectability of the hydrogen molecule (H_2). Although H_2 is an important fuel in combustion, neither LIF nor TDLAS can detect H_2, while spontaneous Raman spectroscopy and CARS have a good sensitivity for H_2. Because of this feature, both methods have often been applied to hydrogen engine combustions. Fluorescence interference from large hydrocarbons such as polycyclic aromatic hydrocarbons (PAHs) does not arise in H_2 combustion, resulting in the preferable condition for spontaneous Raman spectroscopy.

- TDLAS[2.12]–[2.14]

  TDLAS has been applied to concentration and temperature measurements in engine cylinders. Because of the pressure broadening effects, TDLAS has some drawbacks in high-pressure fields, and countermeasures are necessary to compensate for these effects. One of the countermeasures is a wide-range wavelength scanning method to cover the broadened absorption spectra during high-pressure combustion. There are other types of TDLAS applications where TDLAS is used as a sensor of gas temperature and concentration. The measurement device is embedded in a spark plug, and it enables measurements of temperature and species H_2O concentration near the spark plug.

2-D measurements of TDLAS with computer tomography have also been demonstrated in a multicylinder automotive engine. The merit of this method is the fast and continuous imaging of a 2-D measurement section, which is hard to attain by LIF. The smaller size of the laser access ports is also an important feature of TDLAS. Optical access ports, optical fibers, and collimators are embedded in the engine cylinder. It has been demonstrated that this method can detect rapid changes of fuel concentration distribution at a resolution of 3 degrees of crank angle.

2.3.1.3 Exhaust Pipe

Figure 2.7 shows application diagrams of laser diagnostics in exhaust pipes. Exhaust gas management is important for engines and it is necessary to measure exhaust gas compositions with sufficient time resolution to analyze transient engine phenomena.

- LIBS[2.15],[2.16]

  As mentioned above, the fuel-air ratio can be inferred from the elemental analysis of unburned and burned gases. It is useful to know that the equivalence ratio can be inferred from burned gas measurement because the elemental composition does not change during reactions. LIBS can be also used for the elemental analysis of particles, e.g. soot, which contains not only carbon and hydrogen but also metallic elements. Several metallic compositions exist in soot, such as Fe, Mg, Ca, Cu, and Zn, which come from engine wear or lubricant. It is also possible to measure size-classified particles by combining LIBS with a particle classifier.

- TDLAS[2.1],[2.17],[2.18]

  TDLAS has been applied to engine exhaust gas measurements in various approaches. For fast-response measurements, it is possible to achieve the 1 millisecond response time under multispecies and temperature measurement conditions. A sensor unit is attached directly to a flange of the piping, and temperature, H_2O, CO, CO_2, and CH_4 are measured at 1ms response time using a time-division-multiplex method. A clear difference in response is recognized between the measurements by TDLAS and a traditional sampling method. NO_x are important species in engine combustions because they are the main air pollutants in engines. NO_x have been also measured using TDLAS in engine exhaust. NO, NO_2, and N_2O have a strong absorption band in the mid-IR wavelength region, and a quantum cascade laser is mainly used in these applications. An extractive sampling system is often used to apply a quantum cascade laser to engine exhaust measurements.

- LI-TOFMS[2.19]–[2.23]

  In engine exhaust there are several toxic emissions, including NO_x, CO, and soot (or particulate). They have been detected using laser diagnostics such as LIF, TDLAS, and LII. However, none of these methods can detect polycyclic aromatic hydrocarbons (PAHs) at concentration in the range of ppt (10^{-12}). LI-TOFMS can detect PAHs individually with sufficient sensitivity. The sharp rise of PAHs at engine start can be detected using LI-TOFMS. LI-TOFMS has the potential to analyze particle constituents. Particles from diesel engines have been measured using laser desorption/ionization TOFMS. Classified as superfine particles, so-called nanoparticles from diesel engines have also been measured combining LI-TOFMS with differential mobility analyzer (DMA).

2.3.2 Application of Laser Diagnostics to Gas Turbines

Gas turbines have a reasonable scale for laser diagnostics, and similar applications to those of engines have been demonstrated. The main purposes of measurements are the reduction of NO_x and elucidation of phenomena like combustion oscillations. Figure 2.8 shows an application diagram of laser diagnostics to gas turbine combustion analyses.

- LIF[2.2],[2.24]

  LIF has been extensively applied to gas turbines to measure species concentration and temperature. 2-D measurement characteristics are suitable for the elucidation of combustion phenomena in gas turbines. Time-averaged and single-shot OH and NO have been detected to analyze the position and shape of the flame front and the stabilization mechanism. LIF has been applied to both gas turbine burners with a power of 370 kW at 0.3 MPa and with a power of 160 MW at 1.6 MPa.

- Spontaneous Raman spectroscopy and CARS[2.24]

  CARS applications to gas turbines include a temperature measurement in burned gases. Major species concentration such as N_2, O_2, H_2O, CO_2, CH_4, and H_2 and temperature are detected using spontaneous Raman spectroscopy and CARS. It has been demonstrated for gas turbines with a thermal power of 370 kW at 0.3 MPa. Single-shot 1-D Raman measurements have been applied to this burner for quantitative concentration measurements.

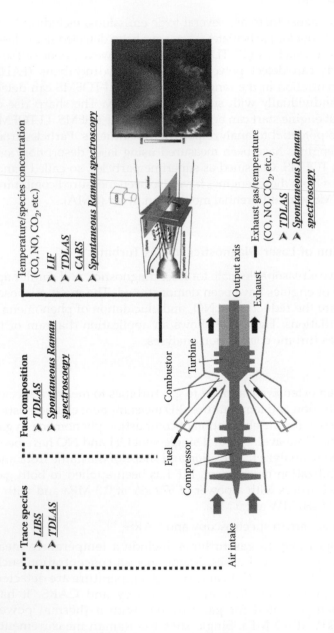

FIGURE 2.8
Application diagrams of laser diagnostics to gas turbines. Gas turbines have a reasonable scale for laser diagnostics and demonstrate similar applications as those of engines. The main purposes of measurements are reduction of NO_x and elucidation of phenomena like combustion oscillations.

- TDLAS[2.25]

    TDLAS has been applied to several types of burners from labora-
    tory scale to large commercial size. These applications extend
    to gas turbines. Temperature and species concentration can be
    measured using TDLAS.

2.3.3 Application of Laser Diagnostics to Large-Scale Burners

In laser diagnostics, many methods have drawbacks in large-scale applications. Beam steering, light collection efficiency, and windows for optical
access become major issues. Therefore several techniques such as a time-of-
flight fluorescence detection method have been employed. TDLAS does not
have a serious disadvantage, and even has advantages under these conditions,
because its signal intensity increases according to the path length. Figure 2.9
shows an application diagram of laser diagnostics to large-scale burners.

- LIF[2.2],[2.26]

    Despite drawbacks of large-area detection, LIF has been applied to
    large-scale burners to develop low NO_x burners. OH and NO
    distributions have been measured in large-scale natural gas
    burners. The wrinkled flame front due to the turbulence of the
    flow can be observed by a single-shot measurement, and the
    area of NO formation is detected in the flame. A unique mea-
    surement technique called a time-of-flight fluorescence detec-
    tion method has also been developed for large-scale combustor
    measurements. The measurement system is equipped with a
    window, and the laser beam is introduced from this window
    into a combustor. The fluorescence signal is detected through
    the same window using a large diameter collection mirror.
    Resolution of 1 m can be achieved using lasers with 5 ns pulse
    width.

- TDLAS[2.2],[2.27],[2.28]

    TDLAS has been applied to several types of burners, from labora-
    tory scale to large commercial size. TDLAS has advantages in
    large-scale measurement conditions because its signal intensity
    increases according to the path length. These applications extend
    to coal-fired boiler burners, incinerator furnaces, and so on.
    Diode lasers with near-infrared (NIR) wavelength are often used
    to measure O_2, H_2O, CO, CO_2, and CH_4. The wavelength conver-
    sion technique extends the available wavelength region to both
    longer and shorter wavelengths, as described in Section 1.4.1.

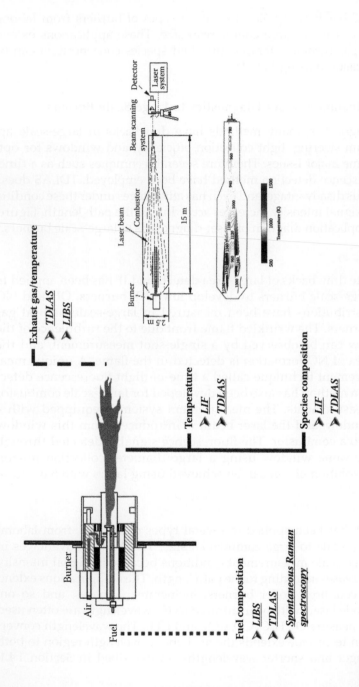

FIGURE 2.9
Application diagrams of laser diagnostics to large-scale burners. In laser diagnostics, many methods have drawbacks in large-scale applications. Beam steering, light collection efficiency, and windows for optical access become major issues. Several techniques have been employed to overcome this drawback. TDLAS does not have a serious disadvantage and even has advantages under these conditions because its signal intensity increases according to the path length.

The 226.8-nm laser output is produced by frequency mixing of a 395-nm external cavity diode laser and a 532-nm laser in a beta-barium-borate crystal. It has been used to measure NO in a 30-kW coal-fired boiler burner. The same method has been applied for the detection of mercury. A 254-nm beam is generated by frequency mixing of a 375-nm single-mode laser and a 784-nm distributed feedback (DFB) and applied to the coal-fired boiler burner.

2.3.4 Application of Laser Diagnostics to Plasma Processes

There are also many studies on plasma diagnostics using laser techniques. Plasma devices like chemical vapor deposition (CVD) are often operated at low pressure and this is preferable for LIF and TDLAS, as shown in Table 2.1.

- LIF[2.29]

  Low-pressure applications such as plasma CVDs are preferable for LIF measurements. One of the important LIF factors is the estimation of quenching rate, and this effect becomes less important at low pressure. Improving and controlling a plasma process is the key factor for the plasma-related devices, and measurements of species concentration and temperature in plasma are necessary for these purposes. In case of SiH_4 plasma CVDs, which are used for solar cell thin films, H, SiH, and SiH_2 are often detected inside plasma. Atomic hydrogen has been detected to better understand the growth process of a diamond film in CVD. It can be shown that H atom concentrations become higher and broader as the number of filaments increases.

- TDLAS[2.30],[2.31]

  CH_4 and C_2H_2 have been monitored for a CVD process using a quantum cascade laser at 7.84 nm. A measurement of HCl has also been demonstrated in a CVD process.

2.3.5 Application of Laser Diagnostics to Plants

Monitoring and advanced control of plants using laser diagnostics has been one of the greatest challenges in its practical applications. Ruggedness and reliability with a reasonable cost are the most important requirements, and Table 2.1 shows that TDLAS is the first candidate for these applications. LIBS and spontaneous Raman spectroscopy are also appropriate to these applications because of their good characteristics in "system cost" and "reliability of laser," as shown in Table 2.1. Figure 2.10 shows an application diagram of laser diagnostics to plants.

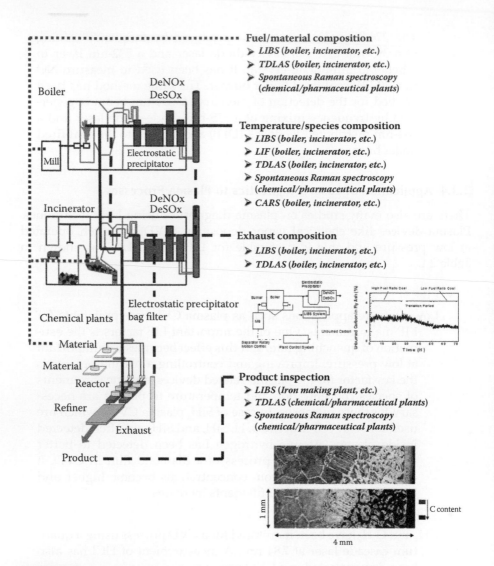

Fuel/material composition
➤ *LIBS (boiler, incinerator, etc.)*
➤ *TDLAS (boiler, incinerator, etc.)*
➤ *Spontaneous Raman spectroscopy (chemical/pharmaceutical plants)*

Temperature/species composition
➤ *LIBS (boiler, incinerator, etc.)*
➤ *LIF (boiler, incinerator, etc.)*
➤ *TDLAS (boiler, incinerator, etc.)*
➤ *Spontaneous Raman spectroscopy (chemical/pharmaceutical plants)*
➤ *CARS (boiler, incinerator, etc.)*

Exhaust composition
➤ *LIBS (boiler, incinerator, etc.)*
➤ *TDLAS (boiler, incinerator, etc.)*

Product inspection
➤ *LIBS (Iron making plant, etc.)*
➤ *TDLAS (chemical/pharmaceutical plants)*
➤ *Spontaneous Raman spectroscopy (chemical/pharmaceutical plants)*

FIGURE 2.10
Application diagrams of laser diagnostics to plants. Monitoring and advanced control of plants using laser diagnostics has been one of the greatest challenges in its practical applications. Ruggedness and reliability with a reasonable cost are the most important requirements. TDLAS is the first candidate for these applications. LIBS and spontaneous Raman spectroscopy are also applicable to these applications because of their good characteristics in cost and reliability.

- TDLAS[2.2],[2.32]–[2.36]

  TDLAS is actively used for process monitoring because of its fast response and noncontact features. Its reasonable cost will enhance this trend. Applications include the aluminum industry, steel-making industry, semiconductor industry, chemical industry, food and pharmaceutical industry, and so on. Currently this process monitoring is conducted mainly by the conventional method, key factors of which are cost and reliability. There are several specific atoms and molecules that are useful in each industry. HF is an important species for the aluminum-making industry because the aluminum smelting process uses alumina (Al_2O_3) and cryolite (Na_3AlF_6), resulting in HF emission. O_2, CO, and CO_2 are important species in many plants, including the steel-making industry and most combustion-related industries. NO_x are also important for the emission control from these processes. In the semiconductor industry, impurities such as H_2O affect plant performance. TDLAS has an excellent sensitivity for H_2O measurement. Systems using TDLAS are simple, and the system can be applicable to many semiconductor industries.

- LIBS[2.27] [2.15]

  LIBS has been actively applied to commercial plants such as iron-making plants, thermal power plants, waste disposal plants, and so on. Many applications have been successfully demonstrated to monitor the plant control factors using LIBS. There are two approaches to apply LIBS to industrial plants. One is the direct monitoring of raw materials or products to optimize the process. The *in situ* characteristics of LIBS are actively utilized in these applications. The other is the detailed measurement of products as a product inspection. In the former application, the main concern is long-term stability and durability of LIBS devices, especially lasers. LIBS usually uses pulsed lasers, and their lifetime usually limits plant applications, especially long-term continuous use for plant monitoring and control. Elemental analyses of metals are one of the latter applications of LIBS, and there have been lots of demonstrations to measure elemental compositions of iron in an iron-making process. An inspection of segregation is one of these examples. LIBS has excellent time and spatial resolutions, and these features are suitable for these applications.

- Spontaneous Raman spectroscopy[2.48]–[2.52]

    Spontaneous Raman spectroscopy has many excellent features for on-line monitoring. Although spontaneous Raman spectroscopy can be applicable to solid, liquid, and gas phase materials, applications to solid and liquid materials are dominant in on-line monitoring. Number densities of solid and liquid materials are much larger than that of gas, and applications of spontaneous Raman spectroscopy to solid and liquid materials can mitigate the drawback of low Raman signal intensities. They include chemical, CVD, and solid pharmaceutical elaboration processes in industry.

2.3.6 Environmental Monitoring and Safety and Security Applications

Environmental monitoring and safety and security applications are growing fields for laser diagnostics. In these cases, ruggedness and reliability become important requirements, and TDLAS and LIBS have been extensively applied in various fields.

- TDLAS[2.53], [2.54]

    Application of TDLAS extends to various fields. In addition to the applications described above, it can be used for environmental monitoring, plant safety, and so on. A detection of carbon isotopes of CO_2 has been also demonstrated in monitoring forest air.

- LIBS[2.55]–[2.61]

    There are many other LIBS applications in various fields. LIBS is intrinsically an elemental analysis method, and it can be appropriate for applications with a need for elemental analyses. One of the major drawbacks of LIBS is the difficulty of quantitative analysis. There are numerous correction methods for LIBS to achieve quantitative information; however, they are usually application dependent and there is no universal method applicable to every LIBS application because of the plasma characteristics produced by a LIBS process. With this in mind, databases for LIBS, which can be usable not only by spectroscopists but designers for various fields, will be important for the advancement of industrial applications.

    Applications of LIBS include all industrial fields, including analyses from food to nuclear materials. Since LIBS is a method for analyzing elemental compositions, industrial fields with a need for elemental analysis can be LIBS applications. As for LIBS applications to foods and water, composition and/or contamination measurements have been demonstrated in flours, wheat, barley, and water. Many elements such as Mg, Al, Cu, Cr, K, Mn, Rb, Cd, and Pb have been measured by LIBS.

2.3.7 Portable Systems

Portability of the system is important for many applications. The portable system is useful for large-area measurements, and it becomes a key technique in many applications.

- LIBS[2.59]–[2.60]

  Portable LIBS systems have been developed and applied to various fields, especially in extraordinary accident and terrorism applications. For example, monitoring of radioactive elements in large areas becomes important in case of leakage of radioactive materials from a nuclear power plant. LIBS can cover these applications, and advances in portable LIBS systems are a key device not only for industries but for safety and security of the people.

- TDLAS[2.61]

  Application of TDLAS extends to various fields. In addition to the applications described above, it can be used for environmental monitoring, plant safety, and so on. A detection of carbon isotopes of CO_2 has been demonstrated as a method of forest air monitoring. It can also be employed as a handheld sensor because the size of diode lasers is much smaller than other lasers.

2.3.8 Life Science and Medical Applications

There is a growing number of applications using laser diagnostics in medical fields. Many of the features of laser diagnostics, such as fast response, excellent spatial resolution, high sensitivity, *in situ* and noncontact detection, are also major requirements of most medical applications. A noncontact or noninvasive measurement is absolutely necessary in many medial applications. Figure 2.11 shows the summary of medical applications using LIF, LIBS, spontaneous Raman spectroscopy, CARS, TDLAS and LI-TOFMS.[2.62]–[2.67] Chapter 8 shows details of these applications.

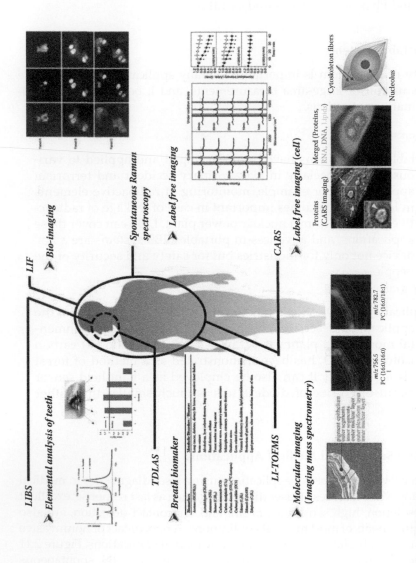

FIGURE 2.11 (SEE COLOR INSERT)
Summary of medical applications using LIF, LIBS, spontaneous Raman spectroscopy, CARS, TDLAS, and LI-TOFMS. Many of the features of laser diagnostics—fast response, excellent spatial resolution, high sensitivity, *in situ* and noncontact detection—are also the major requirements of most of the medical applications. These applications extend from *in vivo* measurements of live cells to biological material analyses.

References

[2.1] M. Yamakage, K. Muta, Y. Deguchi, S. Fukada, T. Iwase, and T. Yoshida, "Development of direct and fast response gas measurement," *SAE Paper* 20081298, 51–59, 2008.

[2.2] Y. Deguchi, M. Noda, Y. Fukuda, Y. Ichinose, Y. Endo, M. Inada, Y. Abe, and S. Iwasaki, "Industrial applications of temperature and species concentration monitoring using laser diagnostics," *Measurement Science and Technology*, 13(10), R103–R115, 2002.

[2.3] W. Kirchweger, R. Haslacher, M. Hallmannsegger, and U. Gerke, "Applications of the LIF method for the diagnostics of the combustion process of gas-IC-engines," *Experiments in Fluids*, 43(2–3), 329–340, 2007.

[2.4] T. Lachaux and M.P.B. Musculus, "In-cylinder unburned hydrocarbon visualization during low-temperature compression-ignition engine combustion using formaldehyde PLIF," *Proceedings of the Combustion Institute*, 31(2), 2921–2929, 2007.

[2.5] M. Löffler, F. Beyrau, and A. Leipertz, "Acetone laser-induced fluorescence behavior for the simultaneous quantification of temperature and residual gas distribution in fired spark-ignition engines," *Applied Optics*, 49(1), 37–49, 2010.

[2.6] T. Kim and J.B. Ghandhi, "Investigation of light load HCCI combustion using formaldehyde planar laser-induced fluorescence," *Proceedings of the Combustion Institute*, 30(2), 2675–2682, 2005.

[2.7] J. Hult, M. Richter, J. Nygren, M. Alden, A. Hultqvist, M. Christensen, and B. Johansson, "Application of a high-repetition-rate laser diagnostic system for single-cycle-resolved imaging in internal combustion engines," *Applied Optics*, 41(24), 5002–5014, 2002.

[2.8] S. Joshi, D.B. Olsen, C. Dumitrescu, P.V. Puzinauskas, and A.P. Yalin, "Laser-induced breakdown spectroscopy for in-cylinder equivalence ratio measurements in laser-ignited natural gas engines," *Applied Spectroscopy*, 63(5), 549–554, 2009.

[2.9] M. Gaft, I. Sapir-Sofer, H. Modiano, and R. Stana, "Laser induced breakdown spectroscopy for bulk minerals online analyses," *Spectrochimica Acta Part B*, 62(12), 1496–1503, 2007.

[2.10] M.C. Weikl, F. Beyrau, and A. Leipertz, "Simultaneous temperature and exhaust-gas recirculation measurements in a homogeneous charge-compression ignition engine by use of pure rotational coherent anti-Stokes Raman spectroscopy," *Applied Optics*, 45(15), 3646–3651, 2006.

[2.11] A. Braeuer and A. Leipertz, "Two-dimensional Raman mole-fraction and temperature measurements for hydrogen–nitrogen mixture analysis," *Applied Optics*, 48(4), B57–B64, 2009.

[2.12] L.A. Kranendonk, J.W. Walewski, T. Kim, and S.T. Sanders, "Wavelength-agile sensor applied for HCCI engine measurements," *Proceedings of the Combustion Institute*, 30(1), 1619–1627, 2005.

[2.13] G.B. Rieker, H. Li, X. Liu, J.T.C. Liu, J.B. Jeffries, R.K. Hanson, M.G. Allen, S.D. Wehe, P.A. Mulhall, H.S. Kindle, A. Kakuho, K.R. Sholes, T. Matsuura, and S. Takatani, "Rapid measurements of temperature and H_2O concentration in IC engines with a sparkplug-mounted diode laser sensor," *Proceedings of the Combustion Institute*, 31(2), 3041–3049, 2007.

[2.14] P. Wright, N. Terzija, J.L. Davidson, S. Garcia-Castillo, C. Garcia-Stewart, S. Pegrum, S. Colbourne, P. Turner, S.D. Crossley, T. Litt, S. Murray, K.B. Ozanyan, and H. McCann, "High-speed chemical species tomography in a multi-cylinder automotive engine," *Chemical Engineering Journal*, 158(1), 2–10, 2010.

[2.15] F. Ferioli, S.G. Buckley, and P.V. Puzinauskas, "Real-time measurement of equivalence ratio using laser-induced breakdown spectroscopy," *International Journal of Engine Research*, 7(6), 447–457, 2006.

[2.16] K. Lombaert, S. Morel, L. LeMoyne, P. Adam, J. Tardieu de Maleissye, and J. Amouroux, "Nondestructive analysis of metallic elements in diesel soot collected on filter: Benefits of laser induced breakdown spectroscopy," Plasma Chemistry and Plasma Processing, 24(1), 41–56, 2004.

[2.17] V.L. Kasyutich, R.J. Holdsworth, and P.A. Martin, "In situ vehicle engine exhaust measurements of nitric oxide with a thermoelectrically cooled, cw DFB quantum cascade laser," *Journal of Physics: Conference Series* 157, 012006, 2009.

[2.18] H. Sumizawa, H. Yamada, and K. Tonokura, "Real-time monitoring of nitric oxide in diesel exhaust gas by mid-infrared cavity ring-down spectroscopy," *Applied Physics B*, 100(4), 925–931, 2010.

[2.19] L. Oudejans, A. Touati, and B.K. Gullett, "Real-Time, on-line characterization of diesel generator air toxic emissions by resonance-enhanced multiphoton ionization time-of-flight mass spectrometry," *Analytical Chemistry*, 76(9), 2517–2524, 2004.

[2.20] B.K. Gulletta, A. Touatib, L. Oudejansb, and S.P. Ryana, "Real-time emission characterization of organic air toxic pollutants during steady state and transient operation of a medium duty diesel engine," *Atmospheric Environment*, 40(22), 4037–4047, 2006.

[2.21] B. Gulletta, A. Touatib, and L. Oudejans, "Use of REMPI–TOFMS for real-time measurement of trace aromatics during operation of aircraft ground equipment," *Atmospheric Environment*, 42(9), 2117–2128, 2008.

[2.22] M. Bente, M. Sklorz, T. Streibel, and Ralf Zimmermann, "Online laser desorption-multiphoton postionization mass spectrometry of individual aerosol particles: Molecular source indicators for particles emitted from different traffic-related and wood combustion sources," *Analytical Chemistry*, 80(23), 8991–9004, 2008.

[2.23] Y. Deguchi, N. Tanaka, M. Tsuzaki, A. Fushimi, S. Kobayashi, and K. Tanabe, "Detection of components in nanoparticles by resonant ionization and laser breakdown time-of flight mass spectroscopy," *Environmental Chemistry*, 5(6), 402–412, 2008.

[2.24] H. Ax, U. Stopper, W. Meier, M. Aigner, and F. Güthe, "Experimental analysis of the combustion behavior of a gas turbine burner by laser measurement techniques," *Journal of Engineering for Gas Turbines and Power*, 132(5), 051503/1–051503/9, 2010.

[2.25] X. Liu, J.B. Jeffries, R.K. Hanson, K.M. Hinckley, and M.A. Woodmansee, "Development of a tunable diode laser sensor for measurements of gas turbine exhaust temperature," *Applied Physics B*, 82(3), 469–478, 2006.

[2.26] M. Versluis, M. Boogaarts, R. Klein-Douwel, B. Thus, W. de Jongh, A. Braam, J.J. Meule, W.L. Meerts, and G. Meijer, "Laser-induced fluorescence imaging in a 100 kW natural gas flame," *Applied Physics B*, 55(2), 164–170, 1992.

[2.27] T.N. Anderson, R.P. Lucht, S. Priyadarsan, K. Annamalai, and J.A. Caton, "In situ measurements of nitric oxide in coal-combustion exhaust using a sensor based on a widely tunable external-cavity GaN diode laser," *Applied Optics*, 46(19), 3946–3957, 2007.

[2.28] J.K. Magnuson, T.N. Anderson, and R.P. Lucht, "Application of a diode-laser-based ultraviolet absorption sensor for in situ measurements of atomic mercury in coal-combustion exhaust," *Energy and Fuels*, 22(5), 3029–3036, 2008.

[2.29] J. Larjo, H. Koivikko, K. Lahtonen, and R. Hernberg, "Two-dimensional atomic hydrogen concentration maps in hot-filament diamond-deposition environment," *Applied Physics B: Lasers and Optics*, 74(6), 583–587, 2002.

[2.30] J. Ma, A. Cheesman, M.N.R. Ashfold, K.G. Hay, S. Wright, N. Langford, G. Duxbury, and Y.A. Mankelevich, "Quantum cascade laser investigations of CH_4 and C_2H_2 interconversion in hydrocarbon/H_2 gas mixtures during microwave plasma enhanced chemical vapor deposition of diamond," *Journal of Applied Physics*, 106(3), 033305/1–033305/15, 2009.

[2.31] V. Hopfe, D.W. Sheel, C.I.M.A. Spee, R. Tell, P. Martin, A. Beil, M. Pemble, R. Weissi, U. Vogth, and W. Graehlerta, "In-situ monitoring for CVD processes," *Thin Solid Films*, 442(1,2), 60–65, 2003.

[2.32] M. Lackner, "Tunable diode laser absorption spectroscopy (TDLAS) in the process industries-a review," *Reviews in Chemical Engineering*, 23(2), 5–147, 2007.

[2.33] H. Gieseler, W.J. Kessler, M. Finson, S.J. Davis, P.A. Mulhall, V. Bons, D.J. Debo, and M.J. Pikal, "Evaluation of tunable diode laser absorption spectroscopy for in-process water vapor mass flux measurements during freeze drying," *Journal of Pharmaceutical Sciences*, 96(7), 1776–1793, 2007.

[2.34] I. Linnerud, P. Kaspersen, and T. Jæger, "Gas monitoring in the process industry using diode laser spectroscopy," *Applied Physics B: Lasers and Optics*, 67(3), 297–305, 1998.

[2.35] J. Ma, A. Cheesman, M.N.R. Ashfold, K.G. Hay, S. Wright, N. Langford, G. Duxbury, and Y.A. Mankelevich, "Quantum cascade laser investigations of CH_4 and C_2H_2 interconversion in hydrocarbon/H_2 gas mixtures during microwave plasma enhanced chemical vapor deposition of diamond," *Journal of Applied Physics*, 106(3), 033305/1–033305/15, 2009.

[2.36] S.C. Schneid, H. Gieseler, W.J. Kessler, and M.J. Pikal, "Non-invasive product temperature determination primary drying using tunable diode laser absorption spectroscopy," *Journal of Pharmaceutical Sciences*, 98(9), 3406–3418, 2009.

[2.37] M. Gaft, I. Sapir-Sofer, H. Modiano, and R. Stana, "Laser induced breakdown spectroscopy for bulk minerals online analyses," *Spectrochimica Acta Part B*, 62(12), 1496–1503, 2007.

[2.38] M. Kurihara, K. Ikeda, Y. Izawa, Y. Deguchi, and H. Tarui, "Optimal boiler control through real-time monitoring of unburned carbon in fly ash by laser-induced breakdown spectroscopy," *Applied Optics*, 42(30), 6159–665, 2003.

[2.39] H.M. Solo-Gabriele, T.G. Townsend, D.W. Hahn, T.M. Moskal, N. Hosein, J. Jambeck, and G. Jacobi, "Evaluation of XRF and LIBS technologies for on-line sorting of CCA-treated wood waste," *Waste Management*, 24(4), 413–424, 2004.

[2.40] T.M. Moskal and D.W. Hahn, "On-line sorting of wood treated with chromated copper arsenate using laser-induced breakdown spectroscopy," *Applied Spectroscopy*, 56(10), 1337–1344, 2002.

[2.41] M.N. Siddiqui, M.A. Gondal, and M.M. Nasr, "Determination of trace metals using laser induced breakdown spectroscopy in insoluble organic materials obtained from pyrolysis of plastics waste," *Bulletin of Environmental Contamination and Toxicology*, 83, 141–145, 2009.

[2.42] J. Anzano, B. Bonilla, B. Montull-Ibor, R.-J. Lasheras, and J. Casas-Gonzalez, "Classifications of plastic polymers based on spectral data analysis with laser induced breakdown spectroscopy," *Journal of Polymer Engineering*, 30(3–4), 177–187, 2010.

[2.43] M.A. Gondal and M.N. Siddiqui, "Identification of different kinds of plastics using laser-induced breakdown spectroscopy for waste management," *Journal of Environmental Science and Health Part A*, 42(13), 1989–1997, 2007.

[2.44] Y. Deguchi, M. Noda, Y. Fukuda, Y. Ichinose, Y. Endo, M. Inada, Y. Abe, and S. Iwasaki, "Industrial applications of temperature and species concentration monitoring using laser diagnostics," *Measurement Science and Technology*, 13(10), R103–R115, 2002.

[2.45] M.M. Tripathi, K.E. Eseller, F.-Y. Yueh, and J.P. Singh, "Multivariate calibration of spectra obtained by laser induced breakdown spectroscopy of plutonium oxide surrogate residues," *Spectrochimica Acta Part B*, 64 (11–12), 1212–1218, 2009.

[2.46] A. Sarkar, V.M. Telmore, D. Alamelu, and S.K. Aggarwal, "Laser induced breakdown spectroscopic quantification of platinum group metals in simulated high level nuclear waste," *Journal of Analytical Atomic Spectrometry*, 24(11), 1545–1550, 2009.

[2.47] T. Hussain and M.A. Gondal, "Detection of toxic metals in waste water from dairy products plant using laser induced breakdown spectroscopy," *Bulletin of Environmental Contamination and Toxicology*, 80(6), 561–565, 2008.

[2.48] N.A. Macleod and P. Matousek, "Emerging non-invasive Raman methods in process control and forensic applications," *Pharmaceutical Research*, 25(10), 2205–2215, 2008.

[2.49] G. Fevotte, "In situ Raman spectroscopy for in-line control of pharmaceutical crystallization and solids elaboration processes: A review," *Chemical Engineering Research and Design*, 85(A7), 906–920, 2007.

[2.50] T.R.M. De Beer, C. Bodson, B. Dejaegher, B. Walczak, P. Vercruysse, A. Burggraeve, A. Lemos, L. Delattreb, Y. Vander Heyden, J.P. Remon, C. Vervaet, and W.R.G. Baeyens, "Raman spectroscopy as a process analytical technology (PAT) tool for the in-line monitoring and understanding of a powder blending process," *Journal of Pharmaceutical and Biomedical Analysis*, 48(3), 772–779, 2008.

[2.51] T.R.M. De Beer, M. Allesø, F. Goethals, A. Coppens, Y. Vander Heyden, H. Lopez de Diego, J. Rantanen, F. Verpoort, C. Vervaet, J.P. Remon, and W.R.G. Baeyens, "Implementation of a process analytical technology system in a freeze-drying process using Raman spectroscopy for in-line process monitoring," *Analytical Chemistry*, 79(21), 7992–8003, 2007.

[2.52] J. Palm, S. Jost, R. Hock, and V. Probst, "Raman spectroscopy for quality control and process optimization of chalcopyrite thin films and devices," *Thin Solid Films*, 515(15), 5913–5916, 2007.

[2.53] S.M. Schaeffer, J.B. Miller, B.H. Vaughn, J.W.C. White, and D.R. Bowling, "Long-term field performance of a tunable diode laser absorption spectrometer for analysis of carbon isotopes of CO_2 in forest air," *Atmospheric Chemistry and Physics*, 8(17), 5263–5277, 2008.

[2.54] F. Zhao, Z. Chen, F. Zhang, R. Li, and J. Zhou, "Ultra-sensitive detection of heavy metal ions in tap water by laser-induced breakdown spectroscopy with the assistance of electrical-deposition," *Analytical Methods*, 2, 408–414, 2010.

[2.55] M. Pouzar, T. Cernohorsky, M. Prusova, P. Prokopcakova, and A. Krejcova, "LIBS analysis of crop plants," *Journal of Analytical Atomic Spectrometry*, 24(7), 953–957, 2009.

[2.56] H.-H. Cho, Y.-J. Kim, Y.-S. Jo, K. Kitagawa, N. Arai, and Y.-I. Lee, "Application of laser-induced breakdown spectrometry for direct determination of trace elements in starch-based flours," *Journal of Analytical Atomic Spectrometry*, 16(6), 622–627, 2001.

[2.57] M.A. Gondal, T. Hussain, Z.H. Yamani, and M.A. Baig, "On-line monitoring of remediation process of chromium polluted soil using LIBS," *Journal of Hazardous Materials*, 163, 1265–1271, 2009.

[2.58] R.S. Harmon, F.C. DeLucia Jr., A. LaPointe, R.J. Winkel Jr., and A.W. Miziolek, "LIBS for landmine detection and discrimination," *Analytical and Bioanalytical Chemistry*, 385(6), 1140–1148, 2006.

[2.59] C.A. Munson, J.L. Gottfried, E. Gibb Snyder, F.C. De Lucia Jr., B. Gullett, and A.W. Miziolek, "Detection of indoor biological hazards using the man-portable laser induced breakdown spectrometer," *Applied Optics*, 47(31), G48–G57, 2008.

[2.60] J.L. Gottfried, F.C. De Lucia Jr, C.A. Munson, and A.W. Miziolek, "Laser-induced breakdown spectroscopy for detection of explosives residues: A review of recent advances, challenges, and future prospects," *Analytical and Bioanalytical Chemistry*, 395(2), 283–300, 2009.

[2.61] M.B. Frish, R.T. Wainner, M.C. Laderer, B.D. Green, and M.G. Allen, "Standoff and miniature chemical vapor detectors based on tunable diode laser absorption spectroscopy," *IEEE Sensors Journal*, 10(3), 639–646, 2010.

[2.62] T. Rosales, V. Georget, D. Malide, A. Smirnov, J. Xu, C. Combs, J. R. Knutson, J.C. Nicolas, and C. A. Royer, "Quantitative detection of the ligand-dependent interaction between the androgen receptor and the co-activator, Tif2, in live cells using two color, two photon fluorescence cross-correlation spectroscopy," *European Biophysics Journal*, 36(2), 153–161, 2007.

[2.63] O. Samek, D.C.S. Beddowsb, H.H. Telle, J. Kaiser, M. Liska, J.O. Caceres, A. G. Urena, "Quantitative laser-induced breakdown spectroscopy analysis of calcified tissue samples," *Spectrochimica Acta Part B*, 56(6), 865–875, 2001.

[2.64] W.-T. Chang, H.-L. Lin, H.-C. Chen, Y.-M. Wu, W.-J. Chen, Y.-T. Lee, and I. Liau, "Real-time molecular assessment on oxidative injury of single cells using Raman spectroscopy," *Journal of Raman Spectroscopy*, 40(9), 1194–1199, 2009.

[2.65] A. Pliss, A. N. Kuzmin, A. V. Kachynski, and P. N. Prasad, "Biophotonic probing of macromolecular transformations during apoptosis," *Proceedings of the National Academy of Sciences of the United States of America*, 107(29), 12771–12776, 2010.

[2.66] C.Wang, and P. Sahay, "Breath Analysis Using Laser Spectroscopic Techniques: Breath Biomarkers, Spectral Fingerprints, and Detection Limits," *Sensors*, 9(10), 8230–8262, 2009.

[2.67] T. Hayasaka, N. Goto-Inoue, Y. Sugiura, N. Zaima, H. Nakanishi, K. Ohishi, S. Nakanishi, T. Naito, R. Taguchi, and M. Setou, "Matrix-assisted laser desorption/ionization quadrupole ion trap time-of-flight (MALDI-QIT-TOF)-based imaging mass spectrometry reveals a layered distribution of phospholipid molecular species in the mouse retina," *Rapid Communications in Mass Spectrometry*, 22(21), 3415-3426, 2008.

3

Laser-Induced Fluorescence

3.1 Principle

In laser-induced fluorescence (LIF), the wavelength corresponding to the electronic energy difference of a molecule is selected as an incident light. Following the absorption of this incident light, the molecule undergoes collision, emission, predissociation, and other processes to transfer into other energy states. The LIF energy transfer process contains many energy levels and energy transfer types, but there are several cases in which a simple two-level model is applicable for the evaluation of the measurement results. The energy transfer process in the two-level model is shown in Figure 3.1. The transfer process is given by the following rate equations:[3.1],[3.2]

$$\frac{dn_1}{dt} = -n_1 W_{12} + n_2(W_{21} + Q_{21} + A_{21})$$ (3.1)

$$\frac{dn_2}{dt} = n_1 W_{12} - n_2(W_{21} + Q_{21} + A_{21} + P_2)$$ (3.2)

where n_1 and n_2 are the number densities in states 1 and 2 respectively, U_L the energy density, A the Einstein A coefficient, Q_{21} the quenching rate, and P_2 the predissociation rate. W_{12} and W_{21} are the stimulated emission and absorption rates, respectively, and they are proportional to the incident laser light intensity. In the case of a steady state, the number density of the measured molecule at the upper excited level n_2 is obtained by the following equation:

$$n_1 W_{12} - n_2(W_{21} + Q_{21} + A_{21} + P_2) = 0$$ (3.3)

When the number density of the lower excited level prior to the laser excitation n^0_1 has a relation $n^0_1 = n_1 + n_2$, which means the loss of number density by predissociation is negligible, Equation (3.3) becomes

$$n_2 = n^0_1 \frac{W_{12}}{W_{12} + W_{21}} \left[1 + \frac{Q_{21} + A_{21} + P_2}{(W_{12} + W_{21})} \right]^{-1} = K_1 n^0_1$$ (3.4)

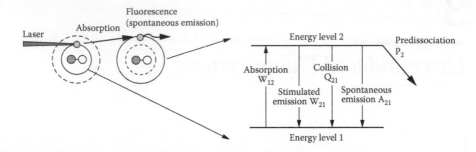

FIGURE 3.1

Two-level energy transfer model. Following the absorption ($B_{12}U_L$) of this incident light, the molecule undergoes collision (Q_{21}), emission (A_{21}), predissociation (P_{21}), and other processes to transfer into other energy states. The emission is known as fluorescence and is used to determine concentration and temperature.

where K_1 is the proportional constant. The fluorescence intensity I is proportional to n_2

$$I = K_2 n_2 A_{21} \tag{3.5}$$

Here, K_2 is a proportional constant. n^0_1 has a relation with the species number density n by the Boltzmann distribution

$$n_1^0 = n \frac{g_1 e^{-E_1/kT}}{\sum_j g_j e^{-E_j/kT}} = \frac{n g_1 e^{-E_1/kT}}{Z} \tag{3.6}$$

where g_i and E_i are the degeneracy and energy of i state, and Z the partition function. Using Equations (3.4) through (3.6), n can be estimated by the fluorescence intensity I

$$n = \frac{n_1^0 Z}{g_1 e^{-E_1/kT}} = \frac{ZI}{K_2 A_{21} \alpha g_1 e^{-E_1/kT}} \tag{3.7}$$

The two different energy states i and j have to be excited to measure temperature

$$n = \frac{n_i^0 Z}{g_i e^{-E_i/kT}} = \frac{ZI_i}{K_2 A_{i,21} \alpha_i g_i e^{-E_i/kT}} \tag{3.8}$$

$$n = \frac{n_j^0 Z}{g_j e^{-E_j/kT}} = \frac{ZI_j}{K_2 A_{j,21} \alpha_j g_j e^{-E_j/kT}} \tag{3.9}$$

Using Equations (3.8) and (3.9), temperature T is given by

$$T = \frac{E_j - E_i}{k \ln\left(\frac{A_{j,21}\alpha_j g_j\, I_i}{A_{i,21}\alpha_i g_i\, I_j}\right)} \tag{3.10}$$

Equation (3.7) is used for concentration measurements and Equation (3.10) for temperature.

There are several difficulties in obtaining quantitative information from the LIF technique, including quenching, broadening of the absorption line, absorption of incident light, and the self-absorption of fluorescence. These effects tend to increase with pressure, and careful consideration is necessary to estimate or eliminate these parameters. The most notable parameter is the quenching rate. The quenching rate Q_{21} of the measurement field is dependent on a quenching rate of each species $Q^q{}_{21}$ in the field and Q_{21} is determined by the following equation.

$$Q_{21} = \sum_q Q^q_{21} = \sum_q n_q v_q \sigma^q \tag{3.11}$$

Here, n_q is the number density of species q, v_q the relative velocity between the measurement species and species q, and σ_q the collisional cross-section between the measurement species and species q. Quenching rates are dependent on number densities of coexisting species as shown in Equation (3.11), and the evaluation of these values is necessary for quantitative measurements.

Other parameters include laser and absorption line shapes, absorption of incident light, and self-absorption of fluorescence. In LIF, the excitation rate is determined by the convolution of the laser light and absorption species line shapes in the wavelength region. Therefore, in addition to the laser light intensity and its line width, it is necessary to consider the absorption line shape of atoms or molecules. The broadening of their absorption lines is a function of the temperature, pressure, and coexistent species. In general, absorption lines become narrow in high-temperature and low-pressure conditions, and broad in low-temperature and high-pressure situations. In the condition of sharp changes in pressure and temperature, care must be exercised to consider the change of excitation efficiency $B_{12}U_L$ according to the change of absorption line shape. Absorption of incident light and self-absorption of fluorescence also affect the excitation rate and signal intensity. In case of optically dense conditions, these effects must be considered carefully because they tend to induce rather large errors in the result.

Several methods have been proposed for the quantitative measurement of LIF. They are briefly summarized here.

1. *Saturated fluorescence.*[3.1] In saturated fluorescence, the laser intensity rises to meet the requirements $P_2 = 0$ and $(W_{12} + W_{21}) \gg Q_{21} + A_{21}$ to remove the effects of quenching effect. It is important to measure the signal in the time domain when the saturation of fluorescence is achieved. Saturated fluorescence has the advantage of the strongest fluorescence intensity; however, planar measurements are difficult in almost all cases because of the need for strong laser-beam intensity.

2. *Predissociation fluorescence.*[3.2] In predissociation fluorescence, the excited energy levels are selected to meet the requirement $P_2 \gg W_{12} + W_{21} + Q_{21} + A_{21}$, which means that the predissociation rate is much faster than that of quenching and that the effect of quenching is removed from Equation (3.1). Fluorescence signals directly from the predissociation level must be measured to ensure the removal of the quenching effect. A predissociation fluorescence example is the OH $A_2\Sigma^+ - X_2\Pi$ (3,0) excitation, and its fluorescence spectrum is shown in Figure 3.2. Since the $A_2\Sigma^+$ $v = 3$ state is a predissociation state, fluorescence lines of (3,1) and (3,2) are signals from the predissociation level. On the other hand, fluorescence lines of (2,1), (1,1), (1,0), and (0,0) are observed as a result of the vibrational relaxation process. In quantitative measurement, the fluorescence lines of (3,1) (3,2) must be measured separately from other fluorescence signals. The depletion of lower excited levels occurs, and the signal intensity becomes dependent on the rotational relaxation rate of lower energy levels if the rate of pumping (intensity of laser light) is too large. Although the predissociation fluorescence method can be applied to two-dimensional quantitative measurements, it has the disadvantage of significantly reduced signal intensity because of the predissociation process.

3. *Cancellation of quenching rate.* For temperature measurements, the quenching effect is canceled if the same excited level is selected in different (often two) absorption lines. As shown in Figure 3.3, not only quenching but vibration-rotation relaxation effects become equivalent and are canceled. If the quenching rates are independent on the excited levels, the effect is also canceled in temperature measurements.

4. *Evaluation of quenching rate.* If the quenching rate is uniform in the measurement area, quantitative measurements are achieved rather easily. It is also possible to directly measure the quenching using time-resolved fluorescence analysis. The time-dependent fluorescence intensity, if $P_{12} = 0$, is proportional to $e^{[-(Q_{21} + A_{21})t]}$ as shown Figure 3.4. Therefore the quenching rate is calculated from the time-decay information of fluorescence signals. The measurement of

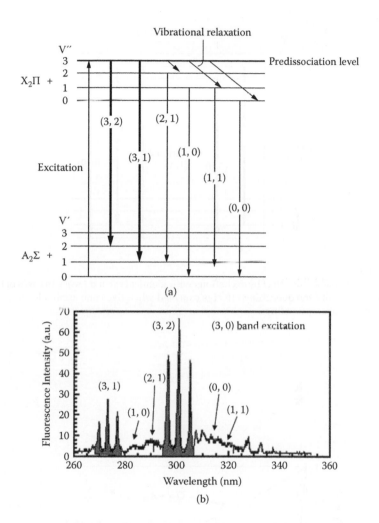

FIGURE 3.2

OH fluorescence by $A_2\Sigma^+ - X_2\Pi$ (3,0) excitation. Fluorescence lines of (3,1) and (3,2) are signals from the predissociation level. On the other hand, fluorescence lines of (2,1), (1,1), (1,0), and (0,0) are observed as a result of the vibrational relaxation process. (a) Energy transfer by $A_2\Sigma^+ - X_2\Pi$ (3,0) excitation. (b) OH fluorescence spectrum.

time-resolved fluorescence intensity often causes complications in measurement schemes for industrial applications.

Although quantitative measurement is an important and challenging problem in LIF, an understanding of planar distributions of molecules or atoms is often enough to get useful information in industrial systems. In this sense a normal LIF method is valid for many applications in industries.

There are several key factors in obtaining accurate signals using LIF. Table 3.1 shows the summary of LIF characteristics. There are several

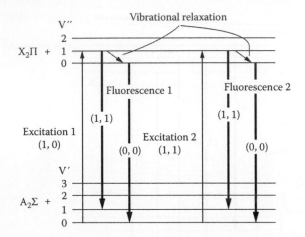

FIGURE 3.3
Cancellation of quenching effect by excitations with identical excited levels. In cases of temperature measurements, the quenching effect is canceled when the same excited level is selected in different two absorption lines.

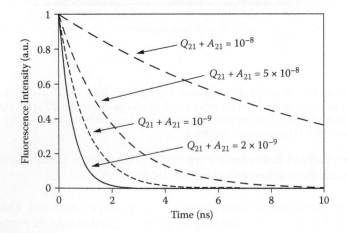

FIGURE 3.4
Time history of fluorescence intensity. Fluorescence intensity decays according to $e^{[-(Q_{21} + A_{21})t]}$. The quenching rate is calculated from the time-decay information of fluorescence signals.

TABLE 3.1

Summary of LIF Characteristics

	Characteristics	Countermeasure
Theoretical treatments	• Rate equations • Quenching	
Temperature	• Boltzmann distribution (lower energy states)	• Theoretical evaluation
Pressure	• Collisional broadening	• Theoretical evaluation
Windows	• Sensitive	• Purge gas • Heating
Calibration	• Difficult (quenching)	• Theoretical evaluation
Noise	• Fluorescence from other molecules • Incident light	• Calibration by tuned and detuned laser lights
Measurement item	• Temperature • Concentration (velocity, pressure)	
Measurement dimension	• 2-D	
Detection limit	• ppb–ppm	
Stability	• Detuning of laser wavelength	• Wavelength monitor

Note: There are several parameters that influence the signal intensity; careful consideration is necessary to estimate or eliminate these parameters. Temperature and pressure effects are important when evaluating LIF signals.

parameters that influence the signal intensity, and careful consideration is necessary to estimate or eliminate these parameters.

1. *Background noise.* One of the LIF features is its strong signal intensity. This method tends to be insusceptible to noise. The most probable noise sources are fluorescence from other molecules and the incident light itself. Fluorescence from other molecules is often reduced by using appropriate filters for both the incident laser light and fluorescence signals. They can also be checked and calibrated by measuring signals using tuned and detuned laser lights. This scheme is illustrated in Figure 3.5. An excellent LIF feature is its ability to separate noises from signals. It is also important to use fluorescence signals that have a wavelength different from the incident laser light if there are scattering bodies like small particles in a measurement area. The wavelength of the incident light is usually shorter than that of fluorescence.

2. *Contamination on measurement windows.* As is often the case with other laser diagnostics, cleanliness of measurement windows has to be maintained. Dirt on measurement windows causes several problems, including attenuation of the incident light and LIF signal

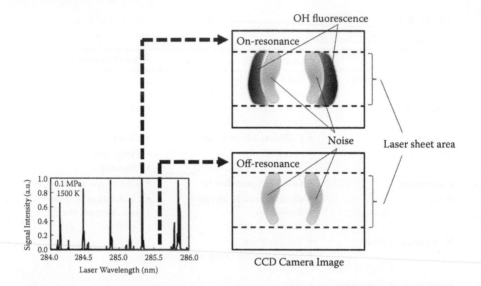

FIGURE 3.5

Noise evaluation using tuned and detuned laser lights. The most probable noise sources are fluorescence from other molecules and the incident light itself. The noise evaluation method using tuned and detuned laser lights is an excellent LIF feature to separate noises from signals.

intensities, and an increase of noise by the scattered incident light. Fluorescence from dirt on windows sometimes causes an increase of noise. If the method is applied in an open field, this problem does not happen, although this is unlikely in industrial applications. It is also important to mention that the termination of the laser light going through a measurement area has to be considered. Allowing the laser light to escape from the measurement area is advisable to maintain the measurement strength.

3. *Pressure effects.* There are two types of pressure effects in LIF because the LIF process consists of absorption and emission phenomena. In the absorption process, the absorption line shape becomes broader at high pressure (see Appendix A). The broad absorption spectra result in the mixing of absorption lines. This tends to induce changes of its excitation efficiency and increase of noise. As for the emission process, the quenching rate depends on pressure. As can be inferred from Equations (3.4) and (3.5), the LIF signal intensity is inversely proportional to the quenching rate and it does not linearly increase with pressure.

4. *Temperature effect on population fraction in energy levels.* The population fraction in each molecular energy level is dependent on temperature based on the Boltzmann equation. The population distribution in energy levels influences the excitation efficiency, which means

fluctuation of the signal intensity. In concentration measurements, it is recommended to choose the excited energy state, which has a uniform population fraction over the temperature range.

5. *Detuning of laser wavelength.* Tunable lasers are often used as a light source in LIF. The laser wavelength has to be tuned to the absorption wavelength of atoms or molecules. The laser wavelength is often stable enough for a LIF measurement in cases where the laser source is placed under fixed temperature and moisture conditions. In several industrial applications, however, lasers are sometimes used in harsh conditions with temperature fluctuations over 10°C. The change of ambient temperature will cause the change (or detuning) of laser wavelength, which results in fluctuation of excitation efficiency.

3.2 Geometric Arrangement and Measurement Species

Many industrial applications employ planar LIF, which means two-dimensional (2-D) image detection of species concentration and/or temperature. A typical geometric arrangement of LIF is shown in Figure 3.6. Tunable lasers such as pulsed dye lasers are often used as a light source to tune the laser wavelength. The laser light is usually formed as a sheet to irradiate the two-dimensional section of a measurement field. The fluorescence signal is collected at right angles to the incoming light and detected by a CCD camera system with appropriate filters. An image-intensified CCD (ICCD) camera is often used to intensify the image and cut out unnecessary light such as flame

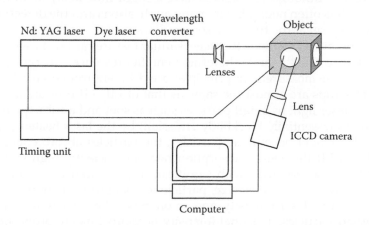

FIGURE 3.6
Typical geometric arrangement of LIF. Main components of a LIF system are a laser and a CCD camera. An ICCD camera is often used to pick out the fast LIF signal from other noise signals.

emissions. Systems for practical applications depend on the conditions, as discussed in Section 3.3.

When LIF is used for practical applications, several key factors were described in Section 3.1. Figure 3.7 shows application procedures that have to be considered in LIF applications. The most important factors for LIF are the tuning of laser wavelength and the evaluation of the quenching rate. Theoretical predictions of LIF spectra are also an important step in order to avoid unnecessary troubles in practical applications.

Figure 3.8 shows theoretical predictions of excitation spectra using Equations (3.4) through (3.6). The OH (1,0) band[3.3],[3.4] at 286 nm is chosen as an example. Notations of the atomic and molecular energy states and their transitions such as "(1,0) band $Q_2(8)$" are often used in the literature of laser diagnostics. These notations are "names" by the rules of spectroscopic theories. These are briefly summarized in Appendix B. The laser wavelength has to be tuned to the atomic or molecular absorption wavelengths if the measured molecule does not have a broad absorption band or the wavelength of a laser accidentally does not coincide with its absorption wavelength. It is worth noting that the identification of the excitation line is important in order to determine the temperature dependence of LIF signal intensities. The quenching rates Q_{21} are assumed to be constant for all excited levels. It can be easily recognized that their structures deeply depend on temperature and pressure. As pressure increases, mixing of excitation lines occurs by the collision broadening (see Appendix C) and excitation spectra become broad. It is desirable to choose the excitation line that has less temperature dependence throughout the considered temperature range using Boltzmann distribution (see Appendix D). The LIF measurement timing chart in case of pulsed lasers as a light source is illustrated in Figure 3.9. Although LIF signals have a slight time delay to an incident laser light, it is often negligible in many applications and the detection timing is usually set at the time of laser irradiation.

There are several methods that have similar features as LIF. Laser-induced incandescence (LII)[3.5] and photofragment fluorescence are typical ones. Although these measurement principles are different from each other, measurement setups are almost the same as that of LIF. LII uses an absorption process of laser light by small particles such as soot and a subsequent emission process (essentially blackbody emission) by the laser-heated particles. When a laser light encounters particles, the particles absorb energy from the light. In LII the energy absorption rate is sufficiently high so that the temperature rises to levels with significant incandescence. The process of LII is dependent on particle size, particle temperature, surrounding ambient temperature, laser intensity, and so on. Since the LII signal is emission from heated particles, its signal intensity is approximately proportional to the particle's volume.

The other method is photofragment fluorescence. The process of photofragment fluorescence[3.5] is almost the same as that of LIF. The only difference

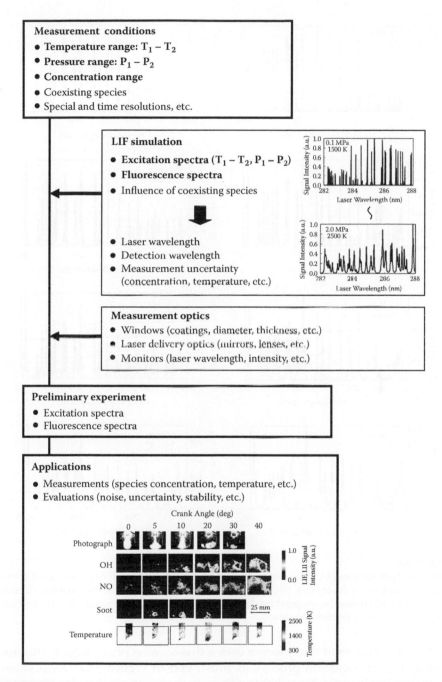

FIGURE 3.7

LIF application procedures. The most important factors for LIF are the tuning of laser wavelength and the evaluation of the quenching rate. Theoretical predictions of LIF spectra are also important to avoid unnecessary troubles in practical applications.

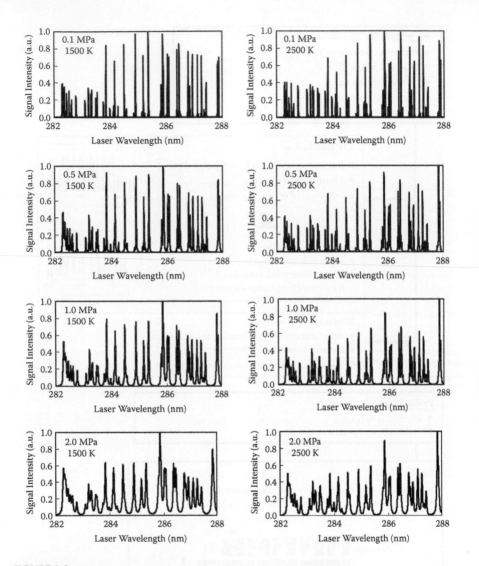

FIGURE 3.8
Theoretical predictions of LIF excitation spectra. The laser wavelength has to be tuned to the atomic or molecular absorption wavelengths, and the identification of the excitation line is important to determine the temperature dependence of LIF signal intensities. Mixing of excitation lines occurs as pressure increases.

is that photofragment fluorescence uses a dissociation phenomenon after (or instead of) an excitation process, as illustrated in Figure 3.10. Dissociation of molecules using a laser light with sufficiently short wavelengths produces electronically excited fragments of atoms or molecules. Fluorescence from these excited fragments, particularly atomic fragments, gives information on the density of "original" molecules. Signal strength as a function of

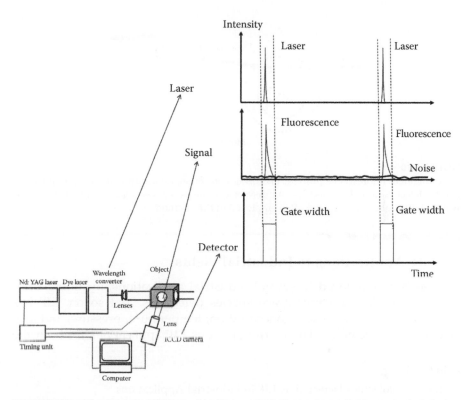

FIGURE 3.9
LIF measurement timing chart. LIF signals have a slight time delay to an incident laser light. It is often negligible in many applications and the detection timing is usually set at the time of laser irradiation.

laser wavelength also indicates differentiation between measured and other molecules.

Typical atoms and molecules detected by LIF in industrial applications are shown in Table 3.2 regarding excitation and detection wavelengths. These species include radicals OH, NO, tracers, and so on.[3.6]–[3.21] Many species have excitation lines at ultraviolet and visible wavelength regions, and these wavelengths are accessible using commercial lasers. Some of the species like O,[3.21] N,[3.21] and H[3.22] atoms have energy transitions at vacuum ultraviolet wavelength—that is, under 200 nm—and these species can be excited using a multiphoton (usually two or three) absorption process. In this case, care must be taken to avoid effects like photofragmentation (or photodissociation) and other unnecessary processes.

The detection limits of these species are often ppm (part per million: 10^{-6}) or lower, and LIF usually shows better characteristics at low pressure because the quenching phenomenon becomes less significant as pressure decreases.

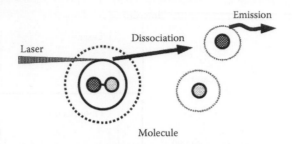

FIGURE 3.10

Photofragment fluorescence process. Photofragment fluorescence uses dissociation phenomena to detect "original" molecules. As dissociation of molecules requires sufficiently short wavelengths light, UV lasers are often employed in this method.

3.3 LIF Applications to Industrial Fields

LIF has been employed in several industrial applications, including combustion and flow analyses, trace species detections, and so on. LIF has been used for visualization, concentration, temperature, pressure, and also velocity measurements. There are numerous LIF applications for both the

TABLE 3.2

Atoms and Molecules Detected by LIF in Industrial Applications

Molecules	Excitation Wavelength (nm)	Detection Wavelength(nm)	References
Hydrogen atom : H	205 (two photon)	656	[3.22]
Oxygen atom : O	226	845	[3.21]
Nitrogen atom : N	211	868	[3.21]
Nitric oxide: NO	193	240–300	[3.8], [3.16],
	225	310	[3.21], [3.33],
	248		
Hydroxyl radical : OH	248		[3.10], [3.11],
	285		[3.12], [3.13],
			[3.16], [3.18],
			[3.19]
			[3.31], [3.32],
			[3.33]
Tracer: Biacetyl	355	400–600	[3.28]
Tracer: Acetone	248–320	350–550	[3.26], [3.29]
Tracer: Formaldehyde	355	350–500	[3.23], [3.25],
			[3.27]
Tracer: Triethylamine	248	260–340	[3.24]

Note: Many species have excitation lines at ultraviolet and visible wavelength regions, and these wavelengths are accessible using commercial lasers. Some of the species like O, N, and H atoms have energy transitions at vacuum ultraviolet wavelength, i.e., under 200 nm, and these species can be excited using a multiphoton (usually two or three) absorption process.

clarification of basic phenomena and the monitoring in industrial processes. LIF often needs tuning of the laser wavelength, which means the necessity of complicated and expensive lasers, and in these cases applications are mainly based on the clarification of basic phenomena in industrial processes; the monitoring and advanced controls of industrial systems have been rather few. Given that LIF can detect radicals such as OH, there is no conventional measurement (chemical analysis) technique to compete with LIF, and the competitors have been other laser diagnostics.

There are two types of approaches for the applications of LIF. One is the measurement method using fluorescent additives, which is often called tracer-LIF,[3.14],[3.23]–[3.27] and the other is the measurement of atoms or molecules that exist naturally in the measurement area. In the former case, molecules such as biacety, acetone, and formaldehyde are used as a tracer. Some of their characteristics are shown in Figure 3.11.[3.28],[3.29] There is usually temperature dependence on the fluorescence signal with each excitation wavelength. In concentration measurements, the excitation wavelength with less temperature dependence is chosen to reduce the variation of signal intensity caused by temperature. However, a set of excitation wavelengths with sufficient temperature dependence is used for temperature measurements. In these cases, tunable lasers are mostly used to tune the wavelength to absorption lines of each atom or molecule.

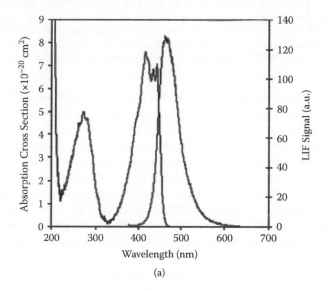

(a)

FIGURE 3.11a
Characteristics of tracer molecules. Absorption (left) and emission (right) spectra (excitation at 355 nm) of gas phase biacetyl. (Reprinted from *Proceedings of the Combustion Institute*, 31(1), J.D. Smith and V. Sick, "Quantitative, dynamic fuel distribution measurements in combustion-related devices using laser-induced fluorescence imaging of biacetyl in iso-octane," 747–755, Copyright 2007, with permission from Elsevier.)

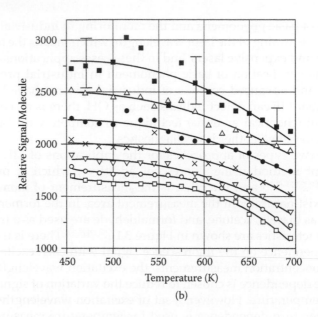

FIGURE 3.11b
Characteristics of tracer molecules. LIF signal intensity of biacetyl excited by 355 nm. Signal versus temperature for seven isobaric conditions: 4 bar (0.4 MPa), 6 bar (0.6 MPa), 8 bar (0.8 MPa), 10 bar (1.0 MPa), 12 bar (1.2 MPa), 14 bar (1.4 MPa), and 16 bar (1.6 MPa). Three sample error bars are shown, for reference, representing the ±10% precision. (Reprinted from *Proceedings of the Combustion Institute*, 31(1), J.D. Smith and V. Sick, "Quantitative, dynamic fuel distribution measurements in combustion-related devices using laser-induced fluorescence imaging of biacetyl in iso-octane," 747–755, Copyright 2007, with permission from Elsevier.)

There are several important devices for these applications, including monitors of laser wavelength and laser sheet intensity distribution. In monitoring laser wavelength, it is important to set the laser wavelength at the right position and to avoid the detuning of the laser wavelength during the measurement period. In combustion analyses, small burners have often been used for this purpose. While monitoring, laser sheet intensity distribution is used to compensate for the effect of laser intensity dependence on fluorescence signals. The laser sheet intensity distribution directly affects the 2-D measurement results and this correction is important to get 2-D results. A 1-D detector with a beam splitter or a cell containing fluorescent species is often used for this purpose. It is also important to remember that this monitor cannot correct the intensity fluctuations caused by measurement windows; cleanliness of windows must be maintained during the measurement period. Schematics of these monitors are shown in Figure 3.12. Fluorescence spectra have sometimes been monitored to confirm the signal quality.

Other measurement techniques such as LII and particle image velocimetry (PIV) are often used along with LIF. LII is intensively employed in engine applications for soot measurements, as soot is an important product

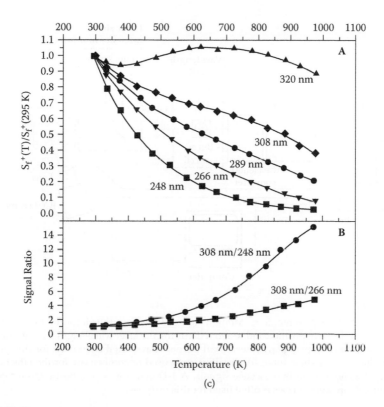

FIGURE 3.11c
Characteristics of tracer molecules. LIF signal intensity of acetone excited by 248, 266, 289, 308, 320 nm. (a) Plots of Sf+ (fluorescence per unit acetone mole fraction) with temperature at $P = 1$, normalized to the room temperature value for each excitation wavelength. (b) Temperature behavior of the fluorescence ratio, generated by dividing one curve Sf^+ by another. (Reprinted with kind permission from Springer Science+Business Media: *Experiments in Fluids*, "Simultaneous imaging of temperature and mole fraction using acetone planar laser-induced fluorescence," 30(1), 93–101, M.C. Thurber and R.K. Hanson, copyright 2007.)

in engine combustion, and LIF cannot be applied to soot measurement. LIF can also be used for velocity measurement using the Doppler effect of light. This method can be applicable to the velocity measurements near or over the velocity of sound. As for velocity measurements, PIV or laser Doppler velocimetry (LDV; see Chapter 8) are much more common in practical applications. It is also worth noting that there have been several applications of pressure measurements using pressure sensitive paints (PSPs).[3.30] This technique uses the fact that fluorescence intensities of PSPs are dependent on pressure, and pressure distribution on the surface with PSPs can be monitored two-dimensionally. Though lasers can be used in this method, many of the applications have used other light sources such as lamps and light-emitting diodes (LEDs) because of their costs.

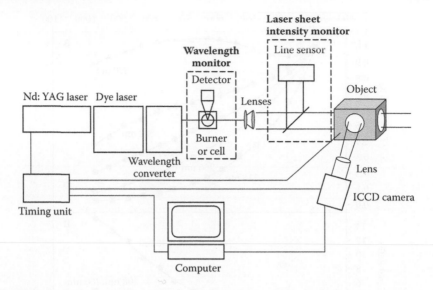

FIGURE 3.12
Schematics of LIF system monitors. Monitoring laser wavelength is important to set the laser wavelength at the right position and to avoid detuning the laser wavelength during a measurement period. In combustion analyses, small burners have often been used for this purpose. The monitor of laser sheet intensity distribution is used to compensate for the effect of laser intensity dependence on fluorescence signals. A 1-D detector with a beam splitter or a cell containing fluorescent species is often used for this purpose.

It is worth noting that there are rapidly growing applications of LIF technologies in the life sciences fields. Visualization of molecules inside live cells using LIF has become a necessary tool to understand live cell functions. These are briefly described in Chapter 8.

3.3.1 Engine Applications

Quite a few applications of LIF have been performed in engine research. One of the reasons is because engines are relatively small in size as industrial systems. The major applications are combustion analyses inside an engine cylinder. In these applications, large temperature and pressure changes arise during a combustion cycle, which makes it difficult to apply LIF for quantitative measurements to these fields. Applications extend to both diesel and gasoline engines. There are also several applications of LIF to jet engines.

3.3.1.1 Measurement Setups

A typical schematic of engine applications is outlined in Figure 3.13.[3.16] The combustion system is equipped with windows for the laser sheet incidence at the side of the combustion chamber and with an observation window

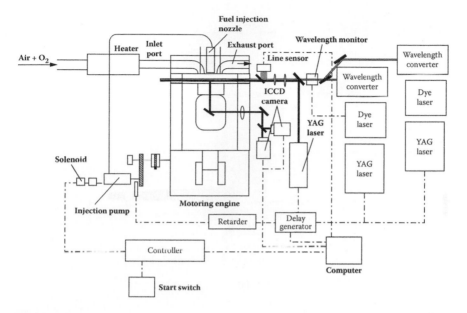

FIGURE 3.13

Schematic of the diesel engine application. Two dye laser systems were used to measure OH, NO, and temperature. Temperature was measured by two-line LIF of NO, and the pressure-induced line broadening effect was evaluated theoretically. The soot was measured using LII with 1064 YAG laser output. (Reprinted from Y. Deguchi, M. Noda, Y. Fukuda, Y. Ichinose, Y. Endo, M. Inada, Y. Abe, and S. Iwasaki, *Measurement Science and Technology*, 13(10), R103, 2002. With permission from the Institute of Physics.)

at the top of the piston to measure signal light in a single-cylinder motoring engine. The fluorescence signal was collected through a UV lens and detected by two CCD cameras, one for concentration measurements and the other for temperature measurements. Synchronization of the system was based on motor rotation, and measurements at any crank angle are achieved by sending a signal with an adequate time delay relative to the motoring rotation to the laser and the CCD cameras. There are several types of measurement systems depending on the aim of measurements, including fuel (or sometimes oxidizer) concentration, burned gas concentration, and temperature measurements.

3.3.1.2 Tracer-LIF Applications

Tracer-LIF is an excellent tool to elucidate the flow and combustion mechanisms in engines. Though it is often used to analyze unburned fuels or oxidizers, it can also be applicable to burned conditions. It is important to know the tracer characteristics in applying this method. Figure 3.14 shows an example of tracer-LIF in a hydrogen engine.[3.24] The measurement results are used to elucidate the jet patterns during injection and the propagation

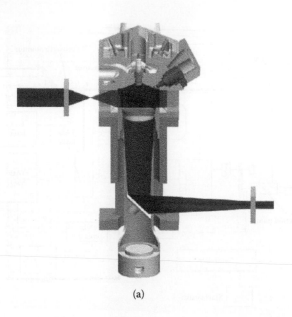

(a)

FIGURE 3.14
Tracer-LIF measurement results in a hydrogen engine. (a) Beam path of LIF measuring system for vertical light sheets. (b) Typical application of quantified tracer-LIF measurements. Comparison of early injection producing a homogeneous charge (left) and late injection resulting in a stratified charge (right). (c) Principle of measuring the flame front propagation. (Reprinted with kind permission from Springer Science+Business Media: *Experiments in Fluids*, "Applications of the LIF method for the diagnostics of the combustion process of gas-IC-engines," 43(2–3), 329–340, W. Kirchweger, R. Haslacher, M. Hallmannsegger, and U. Gerke, copyright 2007.)

of the flame during combustion. The measurement system is almost the same as described above. Triethylamine (TEA) was chosen as the tracer, and a KrF excimer laser at 248 nm was used as an excitation source. Its fluorescence arises between 260 and 340 nm. In Figure 3.14(b), the evolution of the hydrogen jet and its mixing process are clearly recognized, and these results can be used for a direct optimization of mixture formation processes and the validation of computational fluid dynamics (CFD) models. Figure 3.14(c) shows the concept of the flame velocity measurement using this method. Flame velocity can be measured from unburned gas distributions under several conditions, one of which is that the influence of flow on flame propagation is nearly negligible. It is worth noting that in cases of hydrogen combustion, spontaneous Raman spectroscopy can be the other choice for the detection of fuel concentration; its application is shown in Chapter 5.

Figure 3.15 shows another type of tracer-LIF application in engine combustion.[3.25] The results are used to understand the sources of unburned hydrocarbon emissions in a heavy-duty direct injection diesel engine. Formaldehyde (H_2CO) was used as a tracer to visualize the unburned hydrocarbon in engine

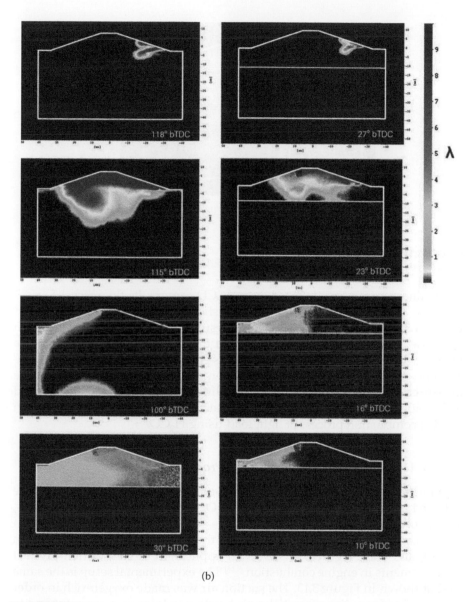

(b)

FIGURE 3.14
(Continued)

combustions. It was selected as a tracer based on chemical kinetic simula-
tions. In applying the tracer-LIF method to reacting fields, it is important to
evaluate the reaction characteristics of the tracer itself. In this application the
spectrograph was added as a detector to confirm that the measured signals
were from the tracer. H_2CO was excited by the Nd:YAG laser at 355 nm, and

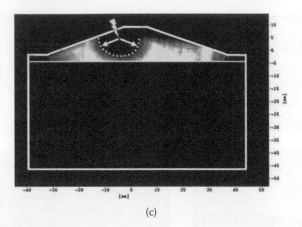

(c)

FIGURE 3.14
(Continued)

H$_2$CO fluorescence was detected by an ICCD camera. Its fluorescence arises between 350 and 500 nm. Figure 3.15(b) shows both the fuel injection and unburned H$_2$CO patterns. Using these results, the authors of this research concluded that mixtures near the injector after the end of injection are too lean to achieve complete combustion, thus contributing to unburned hydrocarbons. The tracer-LIF method has been also used for 2-D temperature measurements in a direct injection gasoline engine using acetone,[3.26] and to investigate quenching of the homogeneous-charge compression-ignition (HCCI) combustion process at light load using H$_2$CO.[3.27]

3.3.1.3 LIF Applications for Combustion Products

Combustion products have often been detected in engines using LIF. The detected molecules include OH, NO, and O$_2$. In addition to these molecules, soot is often measured using LII because soot is an important product in engine combustion. The main purpose of these applications is the reduction of NO$_x$ and soot. Figure 3.16 shows an application of combustion product measurements in engine combustion.[3.16] The experimental setup is the same as that shown in Figure 3.13. The suction air was made oxygen-rich in order to minimize formation of particles obstructive to laser measurement and to maximize formation of NO. From the upper block, it shows the direct flame images taken by a high-speed camera, laser-induced OH fluorescence intensity, laser-induced NO fluorescence intensity, laser-induced soot luminescence intensity, and temperature, respectively. Temperature was measured using two-line LIF of NO and soot using LII. The observed results show that combustion was initiated nearly at the timing of the top dead center, and luminous flame is no longer observed at 30 degrees after top dead center. OH is present outside the region where the flame luminescence is observed,

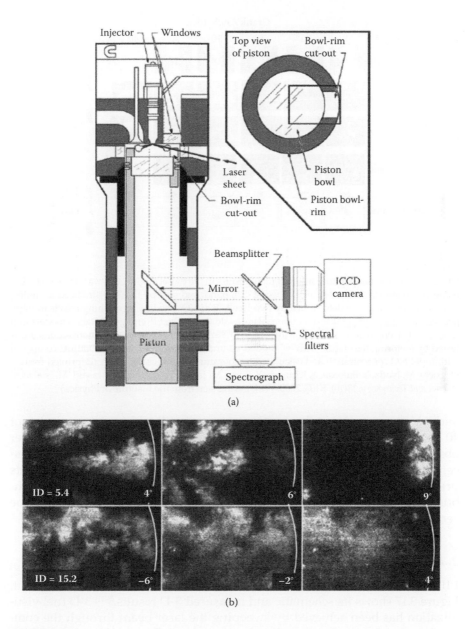

FIGURE 3.15

Tracer-LIF measurement results in a diesel engine. (a) Optical engine and H₂CO PLIF diagnostic. (b) Single-shot PLIF images. Selected representative single-shot PLIF images for the ID = 5.4 CAD condition (top row), and the ID = 15.2 CAD condition (bottom row). The engine crank angle at acquisition is shown in the bottom right of each image. (Reprinted from *Proceedings of the Combustion Institute*, 31(2), T. Lachaux and M.P.B. Musculus, "In-cylinder unburned hydrocarbon visualization during low-temperature compression-ignition engine combustion using formaldehyde PLIF," 2921–2929, Copyright 2007, with permission from Elsevier.)

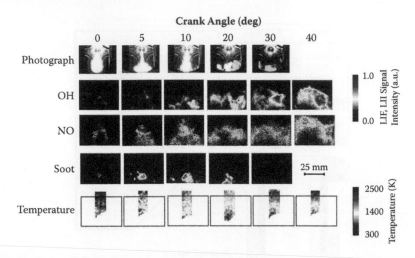

FIGURE 3.16
OH, NO, soot, and temperature measurement results in the diesel engine. OH was detected by LIF using (1,0) band excitation, NO by LIF using (0,0) band, and soot by LII (laser-induced incandescence). Temperature was determined by two-line LIF of NO. The line broadening effects in high-pressure and high-temperature conditions were evaluated by the LIF energy excitation models, and these effects were corrected using the measured pressure time history. Measurement accuracy was tested by motoring the engine with NO-added air and accuracy of 2–5% was obtained compared with input NO concentration and calculated temperature at each crank angle. (Reprinted from Y. Deguchi, M. Noda, Y. Fukuda, Y. Ichinose, Y. Endo, M. Inada, Y. Abe, and S. Iwasaki, *Measurement Science and Technology*, 13(10), R103, 2002. With permission from the Institute of Physics.)

and it is recognized that the reaction process is still taking place at timing of 40 degrees after top dead center where the flame luminescence is no longer observed. The NO distribution is located slightly outside the flame luminescence, in almost the same region as that of OH and high temperature, and its presence in this region tends to increase during the latter period of the combustion process. Soot formation occurs in the fuel-rich region in the flame center and shows a trend similar to that of flame luminescence.

3.3.1.4 3-D LIF Applications

LIF has also been used for the three-dimensional (3-D) measurement. Figure 3.17 shows its schematic and measured 3-D results.[3.31] 3-D fuel visualization has been achieved by sweeping the laser beam through the combustion chamber using a rapidly rotating mirror. Four Nd:YAG lasers, each of which had double laser pulses, were used to make sequences of eight laser pulses, with a time separation of 10 μs each. The total acquisition time of the 3-D data was 70 μs and the separation of each laser sheet was 0.5 mm. Acetone was used as a tracer, and it was excited by 266 nm Nd:YAG radiation. Acetone has a broad absorption band between 225 and 320 nm, and its fluorescence arises between 350 and 550 nm. Figures 3.17(b) and 3.17(c) show

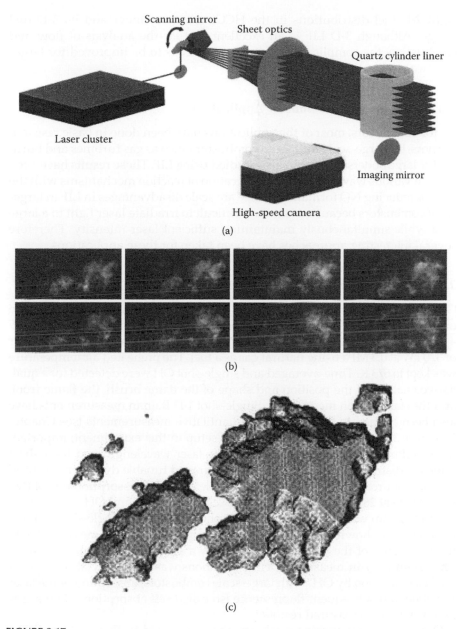

(a)

(b)

(c)

FIGURE 3.17
Schematic of 3-D LIF and 3-D measurement results. (a) Three-dimensional LIF imaging setup.
(b) Three-dimensional fuel tracer LIF. Eight equidistant and parallel 2-D cuts of the fuel distribution in the HCCI engine cylinder are shown. The image separation corresponds to 0.5 mm. The 3-D data were recorded at 6 CAD at top dead center. (c) Three-dimensional fuel iso-concentration surface, calculated from (b). (Reprinted from J. Hult, M. Richter, J. Nygren, M. Alden, A. Hultqvist, M. Christensen, and B. Johansson, *Applied Optics*, 41(24), 5002, 2002. With permission from the Optical Society of America.)

eight 2-D fuel distributions in the HCCI engine cylinder and its 3-D fuel image. Although 3-D LIF is an excellent tool for the analysis of flow and combustion, the complication of its system needs to be improved for larger practical applications.

3.3.2 Large-Scale Combustor Applications

As for combustors, most of the applications have been done in basic research burners, but large-scale industrial combustors such as gas turbines and burners for large boilers have been also studied using LIF. These results have been used in various ways, including clarification of reaction mechanisms with the goal of reducing NO formation. There are scale disadvantages in LIF in large-scale combustors because it is rather difficult to irradiate laser light to a large area while simultaneously maintaining sufficient laser intensity. Therefore several interesting approaches have been taken for these applications.

3.3.2.1 Gas Turbines

Gas turbines have a reasonable scale for LIF measurement and have been extensively studied using laser diagnostics. Figure 3.18 shows one of the LIF applications to gas turbines.[3.32] The burner can produce up to a thermal power of 370 kW at 0.3 MPa using natural gas as a fuel. The preheated air temperature was kept to 673 K. Time-averaged and single-shot OH were detected for a qualitative analysis of the position and shape of the flame brush, the flame front, and the stabilization mechanism. Single-shot 1-D Raman measurements have also been applied to this burner for quantitative measurements (see Chapter 5). Figure 3.18(a) shows the experimental setup in this experiment; important devices illustrated include monitors of the laser wavelength and laser sheet intensity distribution. The Nd:YAG laser pumped tunable dye laser was used as a light source and an ICCD camera as a detector. The absorption line of (1,0) band Q_1 (8) at 283 nm was selected for excitation of OH, and OH fluorescence around 310 nm was observed. In the top photo, mean and single-shot OH distributions are shown in Figure 3.18(c). A strongly wrinkled flame front due to the turbulence of the flow was observed by a single-shot OH measurement. The asymmetry in measured OH distributions was caused by absorption of the laser radiation by OH. With large-scale combustors, absorption of the laser radiation and subsequent fluorescence (so-called self-absorption) often arises and distorts the measured results.

One of other gas turbine applications is illustrated in Figure 3.19.[3.16] The gas turbine combustor has eight main burners (premixed burners) and a pilot burner (diffusion burner) at the center, and 20 combustors of this type can produce 160 MW at a pressure of 1.6 MPa. A quartz cylinder (300 mm in diameter, 5 mm in thickness, and 500 mm in length) was placed behind the nozzle of the combustor, and the air intake was divided by four ducts to observe its flame. Air was introduced to the combustor after heat exchange

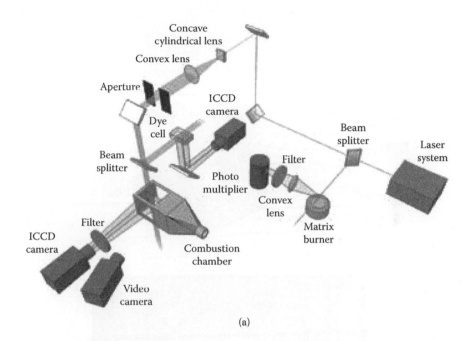

(a)

FIGURE 3.18

LIF application to gas turbines 1. (a) Schematic drawing of the burner and the measurement positions of the different laser diagnostic techniques. (b) Schematic drawing of the experimental setup for OHPLIF measurements. (c) OH-PLIF measurements results. Top: mean OH distribution averaged over 200 single shots. Bottom: single-shot OH-PLIF measurement. (Reprinted from U. Stopper, W. Meier, M. Aigner, and F. Güthe, *Journal of Engineering for Gas Turbines and Power*, 132(5), 051503/1, 2010. With permission from the American Society of Mechanical Engineers.)

with the burned gas. The burners were operated at a pressure of 0.1 MPa to clarify the characteristics of NO formation and turbulent combustion characteristics inside the combustor. The Nd:YAG laser pumped tunable dye laser was used as a light source, and fluorescence signals were detected by an ICCD camera. Time-averaged and single-shot OH and NO distributions were detected in two types of burners: a standard burner and an improved burner for flame stabilization and low NO_x emission. A photograph of the flame at 0.1 MPa is shown in Figure 3.19(b), and time-averaged OH and NO distributions inside the standard burner are shown in Figures 3.19(c) and 3.19(d). OH exists in the mixing area of the pilot gas and the premixed gas of the main burners. OH appears at the outer part of the combustion cylinder. However, NO exists at a high density in the outlet area of the pilot burner; the NO fluorescence intensity gradually drops as the distance from the burner increases. It was found that NO is present mainly at the center of the burner cylinder, and it can be inferred that NO is mainly produced from the reaction of the pilot burner flame. The time-averaged NO distribution inside the improved burner is shown in Figure 3.19(e). The improved burner has a different type of pilot burner to reduce NO_x emission and stabilize its flame.

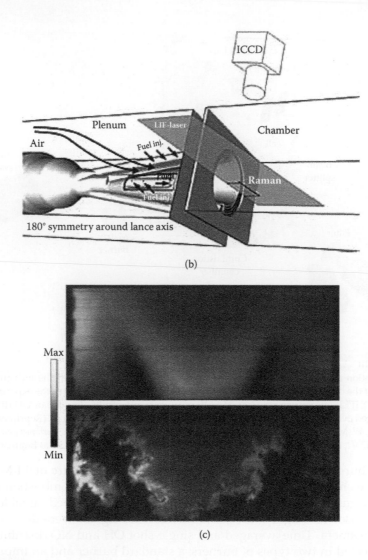

(b)

(c)

FIGURE 3.18
(Continued)

NO fluorescence intensity is also strong near the burner nozzle and exists at the center of the burner cylinder. This is almost the same as in the standard burner, although NO fluorescence intensity in the improved burner extends to the outer part of the burner cylinder near the nozzle. Strong NO formation immediately following combustion at the pilot burner is not seen in the case of the improved burner. The improvement of the pilot burner is intended to provide uniform flame reaction and to reduce partial NO formation.

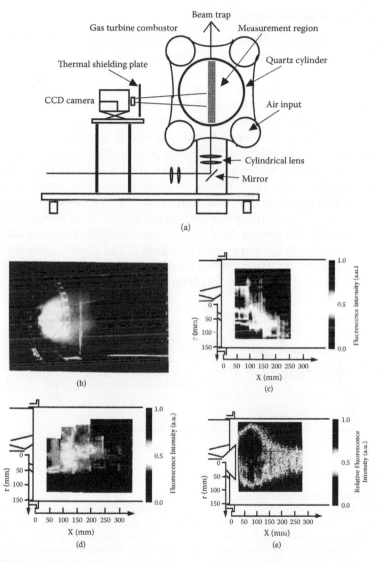

FIGURE 3.19

LIF application to gas turbines 2. The (1,0) band $P_1(6)$ absorption line (284.458 nm) was used in OH measurement and (0,0) band $Q_1(18)$ absorption line (225.786 nm) in NO measurement. Fluorescence corresponding to the (0,0) and (1,1) bands (306 nm–340 nm) was detected in OH, and that corresponding to the (0,2)–(0,6) bands (240 nm–300 nm) was detected in NO. Nonresonant laser rays deviating 0.05 nm from the resonant wavelength were used to remove the background fluorescence in NO measurement. (a) Experimental setup: A dye laser system pumped by an injection-seeded YAG laser was used as the light source. The laser beam was focused by cylindrical lenses to a thickness of 1 mm and a width of 50 mm and was trained on the combustion occurring in the combustor. The fluorescence signal was detected by an ICCD camera system using UV transparent filters. The ICCD camera contains a 286 × 384 photo diode array and can be gated with a minimum gate width of 5 ns. The signals of the ICCD were transferred to a computer and stored for later analysis. (b) Flame photograph of the gas turbine combustor. (c) Time-averaged OH distribution in a normal combustor. (d) Time-averaged OH NO distribution in a normal combustor. (e) Time-averaged OH NO distribution in the improved combustor. (Reprinted from Y. Deguchi, M. Noda, Y. Fukuda, Y. Ichinose, Y. Endo, M. Inada, Y. Abe, and S. Iwasaki, *Measurement Science and Technology*, 13(10), R103, 2002. With permission from the Institute of Physics.)

3.3.2.2 Large-Scale Burners

Despite drawbacks of difficulties in wide-area detection, LIF has also been applied to larger combustors.[3.16],[3.33] The experimental setup for large-scale industrial burner measurement is outlined in Figure 3.20.[3.16] In this application, a unique system has been developed for a large-scale combustor measurement. The measurement system was equipped with a window in the exhaust chamber, and the laser beam was introduced from this window into the combustor. The fluorescence signal was also detected through the same window using a large-diameter collection mirror. Resolution of 1 m inside the combustor can be achieved using a dye laser with 5 ns pulse duration. A photomultiplier tube (PMT) with 0.5 ns response was used as a detector. The laser beam direction was controlled using a computer-controlled three-axis stage. This technique allows greater than 25 m remote measurement ability with 1 m spatial resolution [Figure 3.20(b)]. Two-line LIF of NO was used to measure temperature. LIF signal 1 and signal 2 correspond to the signals induced by two different excitations, that is, two different wavelength laser irradiations at 226 nm [Figure 3.20(c)]. Several burners of this

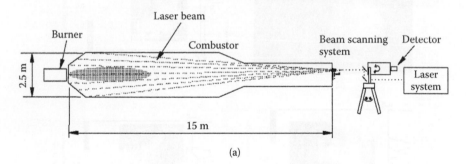

(a)

FIGURE 3.20
LIF application to large-scale burner. Two-line LIF of NO was used to measure temperature. 600 signals (1 min) were accumulated to obtain the temperature information at each measurement line, and seven lines were measured to construct the 2-D temperature map inside the combustor. (a) Experimental apparatus for temperature measurements. (b) Relation between distance and delay time. The speed of light is 3 x 108 m/s, for movement of 0.3 m during 1 ns. The duration between laser irradiation and induced-fluorescence detection time means that the distance between the detector (laser assumed to be located at the same place) and the location of fluorescence emission. A 5-ns laser beam is a 1.5-m bundle of light, and 0.75-m spatial resolution is theoretically achieved considering the back and forth (i.e., fluorescence and laser) light propagation. (c) Typical time history of LIF signals. (d) Temperature measurement result in the large-scale industrial burner. Seven lines were measured to construct the 2-D temperature map inside the combustor. It took 1 min to measure each set of line temperature information, and the measurement (laser beam) direction was controlled automatically using a computer-controlled three-axis stage. Temperature distribution was formed with a peak at about 4 m from the burner along the cylinder axis. Temperature can be seen to decline rapidly toward the exhaust. (Reprinted from Y. Deguchi, M. Noda, Y. Fukuda, Y. Ichinose, Y. Endo, M. Inada, Y. Abe, and S. Iwasaki, *Measurement Science and Technology*, 13(10), R103, 2002. With permission from the Institute of Physics.)

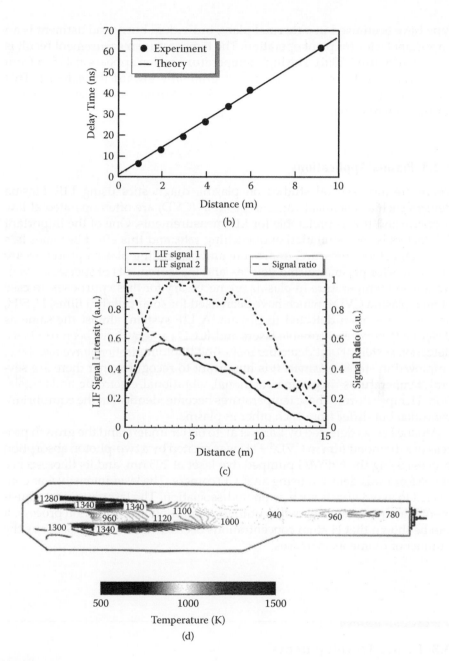

FIGURE 3.20
(Continued)

type have been used in a thermal plant; individual burner adjustment is an important factor for plant operation. The temperature measurement result is shown in Figure 3.20(d). The high-temperature region appears at 4–5 m from the burner and temperature decreases gradually toward the exhaust. This technique can be an excellent tool not only for burner optimization but also for plant monitoring.

3.3.3 Plasma Applications

There are also several studies on plasma diagnostics using LIF. Plasma devices such as chemical vapor deposition (CVD) are often operated at low pressure and this is preferable for LIF measurements. One of the important LIF factors is the estimation of quenching rate, and this effect becomes less important at low pressure. Improving and controlling plasma processes are key factors for plasma-related devices, and measurements of species concentration and temperature in plasma are necessary for these purposes. In case of SiH_4 plasma CVDs, which have been used for solar cell thin films, H, SiH, and SiH_2 are often detected in plasma. A LIF system, almost the same as described above, and tunable lasers and ICCD cameras have been used as a detection system. Point detectors including photomultipliers have also been employed in stable plasmas. It is important to recognize that there are several "temperatures" in plasma: rotational, vibrational, electronic, and translational temperatures. These temperatures become identical in the equilibrium condition but differ from each other in plasma.

Atomic H was detected by Larjo et al. to better understand the growth process of a diamond film in CVD.[3.22] H was excited by a two-photon absorption process using the Nd:YAG pumped dye laser at 205 nm, and its fluorescence at 656.6 nm was detected using an ICCD camera. The laser intensity was controlled to avoid effects such as photodissociation. The laser intensity adjustment is an important step when utilizing a multiphoton excitation process. It can be shown that H atom concentrations become higher and broader as the number of filaments increases.

3.3 Future Developments

There are many other LIF applications in various fields. LIF has been used for visualization, concentration, temperature, pressure, and also velocity measurements. It can be also applied to liquid and solid materials. The typical measurement setup for these applications is shown in Figure 3.18. It is also possible to build portable measurement systems mainly using fixed wavelength lasers. These systems are widely used in various fields. However, LIF applications using tunable lasers have been used mainly for the clarification of basic phenomena in industrial processes. LIF often needs tuning of laser wavelength, which means the necessity of complicated and expensive lasers. Because of their high cost and vulnerability, it is rather difficult to use those laser systems for the control of industrial processes. Tuning of the laser wavelength is not fully automatic in most cases, and users have to tune the laser wavelength manually. In this sense advances in laser development are a key factor for the advancement of future applications. Compared to lasers, detectors are rather robust and easy to use, even for less-skilled engineers.

Three-dimensional detection using LIF is another challenging area in LIF. Although there are several demonstrations of 3D-LIF, measurement systems are still under development and this method has not been widely used even in laboratory experiments. In this case, advances in lasers are also an important step for future development. Two-dimensional and 3-D detections have been an exclusive feature to LIF, but tunable diode laser absorption spectroscopy (TDLAS) using computer tomography is another promising technique in these areas; this latter method is set in motion toward industrial applications. These two methods will soon compete with each other in some applications. As described in Chapter 8, technologies to apply LIF to the life sciences field show rapid growth. Visualization of molecules inside live cells using LIF has also been a promising field for this technology.

References

[3.1]   A.C. Eckbreth, *Laser Diagnostics for Combustion Temperature and Species*, Cambridge, Mass., ABACUS Press, 1988.

[3.2]   E.W. Rothe and P. Andresen, "Application of tunable excimer lasers to combustion diagnostics: A review," *Applied Optics*, 36(18), 3971–4033, 1997.

[3.3]   G.H. Dieke and H.M. Crosswhite, "The ultraviolet bands of OH. Fundamental data," *Journal of Quantitative Spectroscopy & Radiative Transfer*, 2, 97–199, 1962.

[3.4]   W.L. Dimpfl and J.L. Kinsey, "Radiative lifetimes of hydroxyl molecule ($A^2\Sigma$) and Einstein coefficients for the A-X system of hydroxyl and hydroxyl-d molecules," *Journal of Quantitative Spectroscopy & Radiative Transfer*, 21(3), 233–241, 1979.

[3.5]   K. Kohse-Hoinghaus and J.B. Jeffries, *Applied Combustion Diagnostics*, New York, Taylor and Francis, 2002.

[3.6]   D.R. Crosely, "Semiquantitative laser-induced fluorescence in flames," *Combustion and Flame*, 78, 153–167, 1989.

[3.8]   M. Knapp, A. Luczak, V. Beushausen, W. Hentschel, P. Manz, and P. Andresen, "Quantitative in-cylinder NO LIF measurements with a KrF excimer laser applied to a mass-production SI engine fueled with isooctane and regular gasoline," *SAE Technical Paper 970824*, 19–30, 1997.

[3.9]   K. Verbiezen, A.J. Donkerbroek, R.J.H. Klein-Douwel, A.P. van Vliet, P.J.M. Frijters, X.L.J. Seykens, R.S.G. Baert, W.L. Meerts, N.J. Dam, and J.J. ter Meulen, "Diesel combustion: In-cylinder NO concentrations in relation to injection timing," *Combustion and Flame*, 151(1/2), 333–346, 2007.

[3.10] M.G. Allen, K.R. McManus, D.M. Sonnenfroh, and P.H. Paul, "Planar laser-induced-fluorescence imaging measurements of OH and hydrocarbon fuel fragments in high-pressure spray-flame combustion," *Applied Optics*, 34(27), 6287–6300, 1995.

[3.11] K. Kohse-Hoinghaus, U. Meier, and B. Attal-Tretout, "Laser-induced fluorescence study of OH in flat flames of 1–10 bar compared with resonance CARS experiments," *Applied Optics*, 29(10), 1560–1569, 1990.

[3.12] H. Becker, A. Arnold, R. Suntz, P. Monkhouse, J. Wolfrum, R. Maly, and W. Pfister, "Investigation of flame structure and burning behavior in an IC engine simulator by 2D-LIF of hydroxyl radicals," *Applied Physics B*, 50(6), 473478, 1990.

[3.13] A. Arnold, B. Lange, T. Bouche, T. Heitzmann, G. Schiff, W. Ketterle, P. Monkhouse, and J. Wolfrum, "Absolute temperature fields in flames by 2D-LIF of hydroxyl using excimer lasers and CARS spectroscopy," *Berichte der Bunsen-Gesellschaft*, 96(10), 1388–1393, 1992.

[3.14] A.A. Rotunno, M. Winter, G.M. Dobbs, and L.A. Melton, "Direct calibration procedures for exciplex-based vapor/liquid visualization of fuel sprays," *Combustion Science and Technology*, 71(4–6), 247–261, 1990.

[3.16] Y. Deguchi, M. Noda, Y. Fukuda, Y. Ichinose, Y. Endo, M. Inada, Y. Abe, and S. Iwasaki, "Industrial applications of temperature and species concentration monitoring using laser diagnostics," *Measurement Science and Technology*, 13(10), R103–R115, 2002.

[3.17] S. Einecke, C. Schultz, and V. Sick, "Measurement of temperature, fuel concentration and equivalence ratio fields using tracer LIF in IC engine combustion," *Applied Physics B*, 71(5), 717–723, 2000.

[3.18] P. Pixner, R. Schiessl, A. Dreizler, and U. Maas, "Experimental determination of pdfs of OH radicals in IC engines using calibrated laser-induced fluorescence as a basis for modelling the end-phase of engine combustion," *Combustion Science and Technology*, 158, 485–509, 2000.

[3.19] S.R. Engel, P. Koch, A. Braeuer, and A. Leipertz, "Simultaneous laser-induced fluorescence and Raman imaging inside a hydrogen engine," *Applied Optics*, 48(35), 6643–6650, 2009.

[3.20] E.A. Brinkman, G.A. Raiche, M.S. Brown, and J.B. Jeffries, "Optical diagnostics for temperature measurement in a d.c. arcjet reactor used for diamond deposition," *Applied Physics B*, 64(6), 689–697, 1997.

[3.21] A. Broc, S.D. Benedictis, and G. Dilecce, "LIF investigations on NO, O and N in a supersonic N2/O2/NO RF plasma jet," *Plasma Sources Science & Technology*, 13(3), 504–514, 2004.

[3.22] J. Larjo, H. Koivikko, K. Lahtonen, and R. Hernberg, "Two-dimensional atomic hydrogen concentration maps in hot-filament diamond-deposition environment," *Applied Physics B: Lasers and Optics*, 74(6), 583–587, 2002.

[3.23] T. Kim, J.B. Ghandhi, "Investigation of light load HCCI combustion using formaldehyde planar laser-induced fluorescence," *Proceedings of the Combustion Institute*, 30(2), 2675–2682, 2005.

[3.24] W. Kirchweger, R. Haslacher, M. Hallmannsegger, and U. Gerke, "Applications of the LIF method for the diagnostics of the combustion process of gas-IC-engines," *Experiments in Fluids*, 43(2–3), 329–340, 2007.

[3.25] T. Lachaux and M.P.B. Musculus, "In-cylinder unburned hydrocarbon visualization during low-temperature compression-ignition engine combustion using formaldehyde PLIF," *Proceedings of the Combustion Institute*, 31(2), 2921–2929, 2007.

[3.26] M. Löffler, F. Beyrau, and A. Leipertz, "Acetone laser-induced fluorescence behavior for the simultaneous quantification of temperature and residual gas distribution in fired spark-ignition engines," *Applied Optics*, 49(1), 37–49, 2010.

[3.27] T. Kim and J.B. Ghandhi, "Investigation of light load HCCI combustion using formaldehyde planar laser-induced fluorescence," *Proceedings of the Combustion Institute*, 30(2), 2675–2682, 2005.

[3.28] J.D. Smith and V. Sick, "Quantitative, dynamic fuel distribution measurements in combustion-related devices using laser-induced fluorescence imaging of biacetyl in iso-octane," *Proceedings of the Combustion Institute*, 31(1), 747–755, 2007.

[3.29] M.C. Thurber and R.K. Hanson, "Simultaneous imaging of temperature and mole fraction using acetone planar laser-induced fluorescence," *Experiments in Fluids*, 30(1), 93–101, 2001.

[3.30] J.W. Gregory, K. Asai, M. Kameda, T. Liu, and J.P. Sullivan, "A review of pressure-sensitive paint for high-speed and unsteady aerodynamics," *Proceedings of the Institution of Mechanical Engineers G*, 222(2), 249–290, 2008.

[3.31] J. Hult, M. Richter, J. Nygren, M. Alden, A. Hultqvist, M. Christensen, and B. Johansson, "Application of a high-repetition-rate laser diagnostic system for single-cycle-resolved imaging in internal combustion engines," *Applied Optics*, 41(24), 5002–5014, 2002.

[3.32] U. Stopper, W. Meier, M. Aigner, and F. Güthe, "Experimental analysis of the combustion behavior of a gas turbine burner by laser measurement techniques," *Journal of Engineering for Gas Turbines and Power*, 132(5), 051503/1–051503/9, 2010.

[3.33] M. Versluis, M. Boogaarts, R. Klein-Douwel, B. Thus, W. de Jongh, A. Braam, J.J. Meulen, W.L. Meerts, and G. Meijer, "Laser-induced fluorescence imaging in a 100 kW natural gas flame," *Applied Physics B*, 55(2), 164–170, 1992.

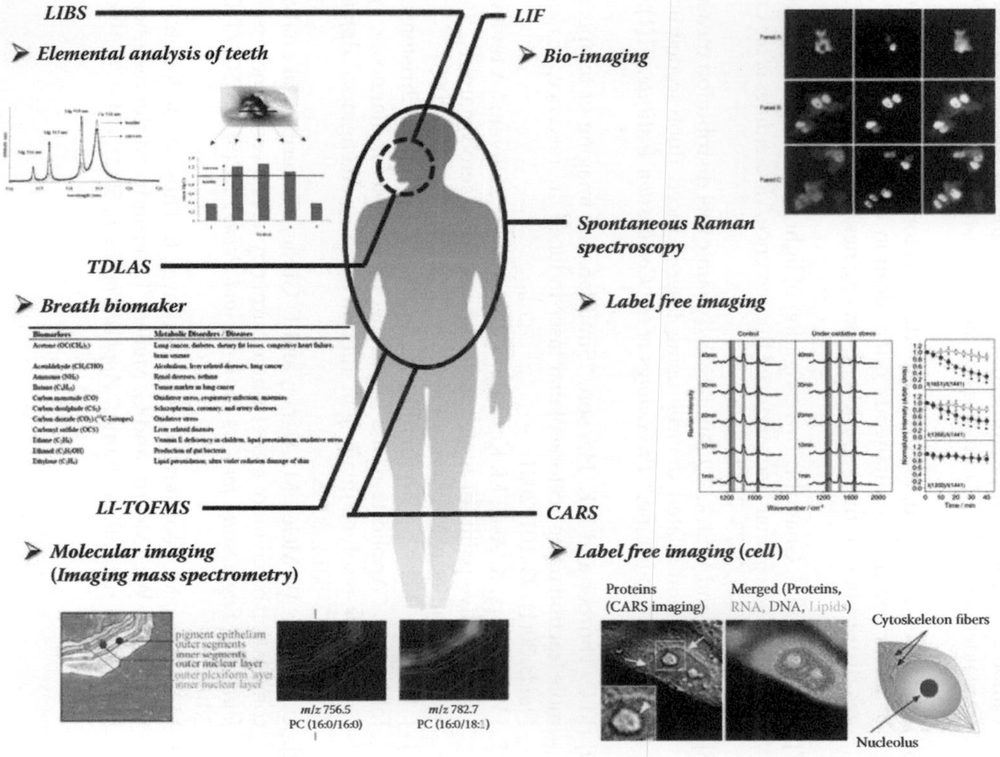

COLOR FIGURE 2.11

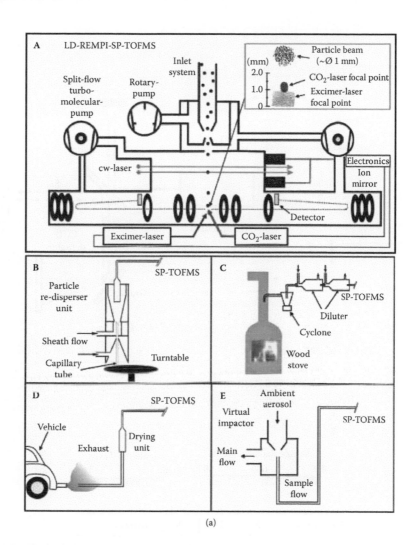

(a)

COLOR FIGURE 7.10a

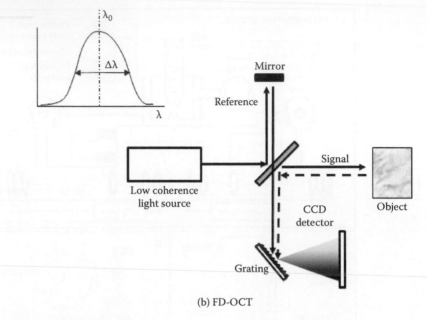

(b) FD-OCT

COLOR FIGURE 8.2b

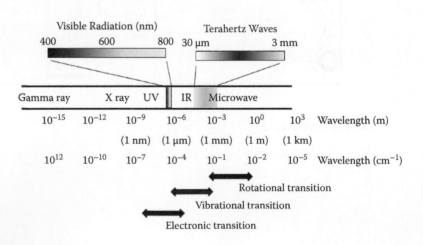

COLOR FIGURE 8.5

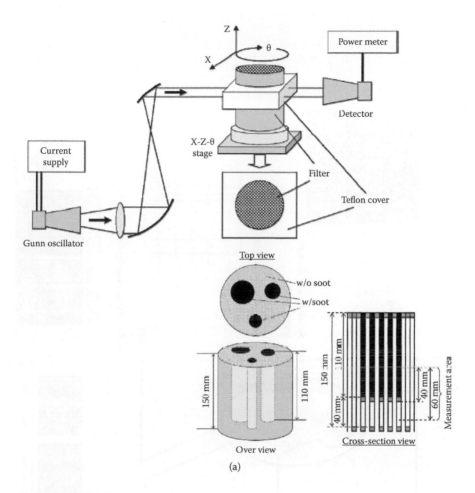

COLOR FIGURE 8.8a

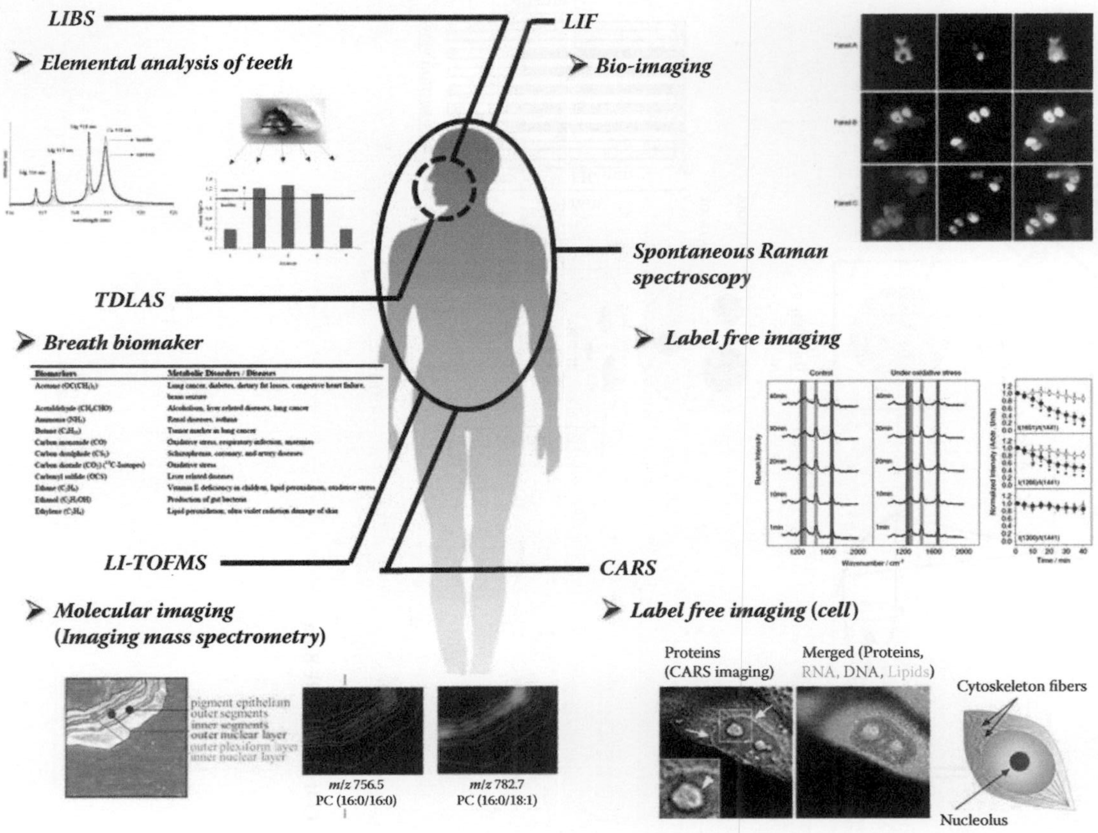

LIBS
➤ Elemental analysis of teeth

LIF
➤ Bio-imaging

TDLAS
➤ Breath biomaker

Biomarkers	Metabolic Disorders/Diseases
Acetone (OC(CH₃)₂)	Lung cancer, diabetes, dietary fat losses, congestive heart failure
Acetaldehyde (C₃H₄CHO)	Liver cancer, liver-related diseases, lung cancer
Ammonia (NH₃)	Renal disease
Butane (C₄H₁₀)	Tissue marker as lung cancer
Carbon monoxide (CO)	Oxidative stress, respiratory infection, anemia
Carbon sulphide (CS₂)	Schizophrenia, coronary and artery diseases
Carbon dioxide (CO₂)/(¹³CO₂ isotopes)	Oxidative stress
Cumol carbon sulfide (COS)	Liver diseases
Ethane (C₂H₆)	Vitamin E deficiency in children, lipid peroxidation, oxidative stress
Ethanol (C₂H₅OH)	Production of gut bacteria
Ethylene (C₂H₄)	Lipid peroxidation, skin cancer radiation damage of skin

Spontaneous Raman spectroscopy
➤ Label free imaging

CARS
➤ Label free imaging (cell)

Proteins (CARS imaging)   Merged (Proteins, RNA, DNA, Lipids)

Cytoskeleton fibers

Nucleolus

LI-TOFMS
➤ Molecular imaging (Imaging mass spectrometry)

pigment epithelium
outer segments
inner segments
outer nuclear layer
outer plexiform layer
inner nuclear layer

m/z 756.5 PC (16:0/16:0)

m/z 782.7 PC (16:0/18:1)

COLOR FIGURE 8.9

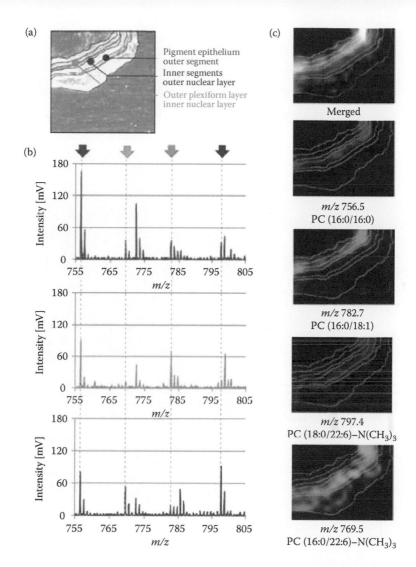

(a)

Pigment epithelium
outer segment
Inner segments
outer nuclear layer
Outer plexiform layer
inner nuclear layer

(b)

(c)

Merged

m/z 756.5
PC (16:0/16:0)

m/z 782.7
PC (16:0/18:1)

m/z 797.4
PC (18:0/22:6)–N(CH$_3$)$_3$

m/z 769.5
PC (16:0/22:6)–N(CH$_3$)$_3$

COLOR FIGURE 8.15

4

Laser-Induced Breakdown Spectroscopy

4.1 Principle

LIBS is the acronym for "laser-induced breakdown spectroscopy," and it is also called LIPS (laser-induced plasma spectroscopy) and other related names.[4.1] The principle behind LIBS is illustrated in Figure 4.1. In the LIBS process, a laser beam is focused onto a small area, producing hot plasma. The material contained in plasma is atomized, and light is released corresponding to a unique wavelength for each element. Despite the fact that the processes involved are complex, the emission intensity from the atomized species can be described by the following equation with the assumption of a uniform plasma temperature:

$$I_{(i)} = n_{(i)} \sum_j \left\{ K_{(i),j} g_{(i),j} \exp\left(-\frac{E_{(i),j}}{kT} \right) \right\} \tag{4.1}$$

In the above expression, $I_{(i)}$ is the emission intensity of species i, $n_{(i)}$ the concentration of species i, $K_{(i),j}$ a variable that includes the Einstein A coefficient from the upper energy level j, $g_{(i),j}$ the statistical weight of species i at the upper energy level j, $E_{(i),j}$ the upper energy level of species i, k the Boltzmann constant, and T the plasma temperature. In Equation (4.1), there are several factors that affect the emission intensity $I_{(i)}$. These include plasma temperature, plasma nonuniformity, and matrix effects. The appropriate correction factors must be included in $K_{(i),j}$ to obtain quantitative results.

A calibration of the LIBS signal is necessary for quantitative analyses. The signal intensity of LIBS depends on several factors, including plasma temperature, and its signal essentially fluctuates. Because of these intrinsic characteristics of LIBS, the signal intensity ratio has often been used for elemental composition analyses. This procedure can eliminate some of the fluctuation factors. It is, however, necessary to cancel out the plasma temperature effects

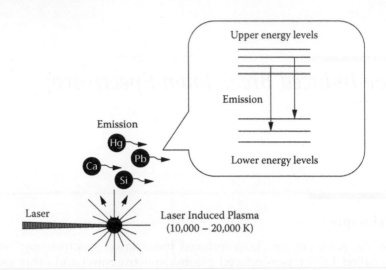

FIGURE 4.1

Principle of LIBS. Plasma induced by a laser beam instantaneously reaches the 10,000–20,000 K temperature regions. The plasma first emits strong continuous noise, with atomic emissions appearing after a specific time delay. The necessary time delay is dependent on the upper energy of the measured element and plasma conditions.

from the signal intensity ratio. According to Equation (4.1), the emission intensities of species i_1 at the upper energy levels $j1$ and $j2$ are given by

$$I_{(i),j1} = n_{(i)} K_{(i),j1} g_{(i),j1} \exp\left(-\frac{E_{(i),j1}}{kT}\right) \tag{4.2}$$

$$I_{j(1),j2} = n_{(i)} K_{(i),j2} g_{(i),j2} \exp\left(-\frac{E_{(i),j2}}{kT}\right) \tag{4.3}$$

From Equations (4.2) and (4.3), the plasma temperature T has the following relation with the emission intensity ratio $I_{(i),j1}/I_{(i),j2}$:

$$\exp\left(-\frac{1}{kT}\right) = \left(\frac{I_{(i),j1}}{I_{(i),j2}} \frac{K_{(i),j2}}{K_{(i),j1}} \frac{g_{(i),j2}}{g_{(i),j1}}\right)^{\frac{1}{(E_{(i),j1}-E_{(i),j2})}} \propto \left(\frac{I_{(i),j1}}{I_{(i),j2}}\right)^{b_{(i)}} \tag{4.4}$$

where $b_{(i)}$ is a plasma temperature correction factor. It is inferred from Equation (4.4) that the effect of plasma temperature is corrected by the emission intensity ratio $I_{(i),j1}/I_{(i),j2}$. It means that different emission lines from the same species can be selected to cancel the plasma temperature dependence of the emission intensity.

In LIBS one of the emissions in Equation (4.1) is often selected for the elemental composition measurement. In this case $n_{(i)}$ can be recast in the form

$$n_{(i)} = \frac{I_{(i),j}}{K_{(i),j} g_{(i),j} \exp\left(-\frac{E_{(i),j}}{kT}\right)} \tag{4.5}$$

There are several forms in which plasma correction terms can be put into Equation (4.5). With the matrix effects considered, one of the forms may have the following relation:

$$n_{(i)} = K_{(i)} \left(I_{(i),j}\right)^{b_0} \prod_{i1} \left(\frac{I_{(i1),j1}}{I_{(i1),j2}}\right)^{b_{(i1)}} \tag{4.6}$$

where $K_{(i)}$ and b_0 are correction factors of species i, and $b_{(i1)}$ is the temperature correction factor in the emission pair of $I_{(i),j1}$ and $I_{(i),j2}$.

A typical correction curve of the plasma temperature is shown in Figure 4.2.[4.2] In this case, the two Mg lines, shown Mg_1 and Mg_2 in the figure, are used for the plasma temperature correction. It is clear from Figure 4.2(b) that the ratios are influenced by the plasma condition, which is directly related in the ratio Mg_1/Mg_2. The results are shown in Figure 4.2(c). By applying the temperature correction scheme, the I_{si}/I_{Al} ratios become constant even for different plasma conditions, that is, Mg_1/Mg_2. These correction factors are dependent on the pressure, emission detection delay time, detection time width, laser intensity, and other factors. Remember that LIBS employs the plasma generation process using pulsed laser energy, which is not in the equilibrium condition at all. There are also matrix effects in LIBS signal intensities that cannot be predicted using Equation (4.4). Therefore, the calibration method described above is an empirical one, and the correction factors are valid only in the specific experimental conditions. These factors have to be determined in each measurement condition.

There are several important factors to get quantitative information using LIBS. Table 4.1 summarizes LIBS characteristics. This method is experimentally simple. However, it is very difficult to solve the LIBS process theoretically because it contains laser-material interactions, rapid temperature changes over 10,000 K in a nano- or picosecond time scale, and plasma cooling phenomena that include the recombination process of ions, electrons, and neutrals. Therefore, choosing appropriate experimental parameters is important to make the theoretical treatment applicable for quantitative measurements.

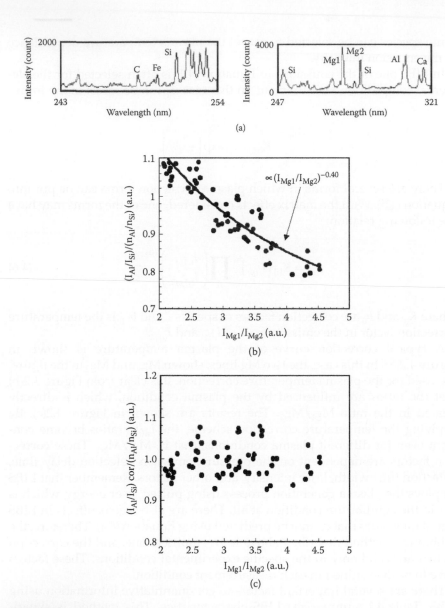

FIGURE 4.2

Typical correction curve of the plasma temperature. Data were collected using several different fly ash samples and different plasma conditions. The effect of the species concentration in different fly ash samples was canceled by dividing I_C/I_{Si} by n_C/n_{Si}. It is clear that the ratio I_C/I_{Si} is influenced by the condition of plasma, which is directly related in the ratio Mg_1/Mg_2. By applying the temperature correction scheme, these ratios become constant even for different plasma conditions. (a) LIBS spectrum of fly ash from a pulverized coal power plant. (b) Correction factor for the I_C/I_{Si} ratio. (c) Corrected Results for the I_C/I_{Si} ratio. (Reprinted from *Spectrochimica Acta B*, 57(4), M. Noda, Y. Deguchi, S. Iwasaki, and N. Yoshikawa, "Detection of carbon content in a high-temperature and high-pressure environment using laser-induced breakdown spectroscopy," 701–709, Copyright 2002, with permission from Elsevier.)

TABLE 4.1

Summary of LIBS Characteristics

	Characteristics	Countermeasure
Theoretical treatments	• Boltzmann distribution • Emission	
Temperature	• Boltzmann distribution (upper energy states)	• Theoretical evaluation
Pressure	• Collision (plasma cooling)	• Delay time and gate width
Windows	• Insensitive	• Purge gas • Heating
Calibration	• Empirical	• Calibration curve • Plasma temperature correction
Noise	• Plasma emission (black-body-like emission)	• Delay time and gate width
Measurement item	• Element	
Measurement dimension	• Point	• Scanning
Detection limit	• ppb–ppm	• Double pulse • LIBS + LIF • LIBS + TOFMS
Stability	• Laser (long term)	

Note: LIBS is experimentally simple. However, it is very difficult to solve the LIBS process theoretically because it contains laser-material interactions and rapid temperature changes over 10,000 K in a nano- or picosecond timescale. Choosing appropriate experimental parameters is important to make the theoretical treatment applicable for quantitative measurements.

1. *Background noise.* The main background noise of LIBS is the black-body-like emission from plasma itself. Typical LIBS signals according to the time delay from the laser input are shown in Figure 4.3. Atomic emissions appear after a certain time delay, which means that LIBS signals appear during the plasma cooling process. Therefore, it is important to choose the appropriate delay time and gate width. Although it is possible to get the LIBS signal without any gating, the detection limit becomes much lower than that with appropriate gate width.

2. *Stability of plasma.* As mentioned above, the signal intensity of LIBS depends on several factors such as plasma temperature, and fluctuation of its signal is an intrinsic characteristic in LIBS. Use of the signal intensity ratio and plasma temperature correction can reduce this fluctuation to some extent. Since the correction technique can be applicable within limited experimental conditions, experimental parameters such as a laser power and measured material conditions have to be controlled within defined values. In the solid material analysis, it is also important to measure the target with a fresh surface in each laser shot by scanning a laser pass or moving the target.

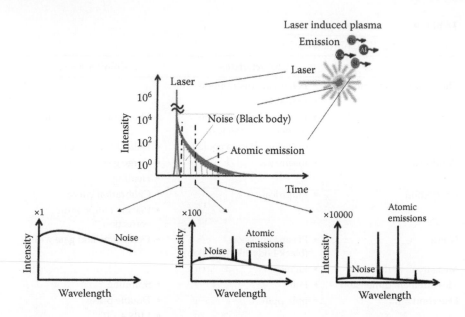

FIGURE 4.3
LIBS signals according to the time delay from the laser input. The main background noise of LIBS is the black-body-like emission from plasma itself. Atomic emissions appear after a certain time delay, which means that LIBS signals appear during the plasma cooling process.

3. *Nonuniformity of plasma.* The laser-induced plasma has its structure in it. In the plasma generation process, the plasma structure depends largely on the laser density pattern and measured material conditions. After the plasma generation, the induced plasma expands rapidly into an ambient environment. Therefore, the core of the plasma is usually hotter than its edge, and the LIBS signal depends on the measurement area across the plasma. In this sense the correction method mentioned above also includes the effects of plasma nonuniformity in it.

4. *Matrix effects.* Matrix effects are the combined effects of all components other than the measured species. Changes of these components may cause alteration of LIBS signal intensity even if the number density of the measured species is the same throughout the measurement period. Matrix effects intrinsically exist in LIBS, and they are usually corrected by experimental calibration.

5. *Dirt on measurement windows.* In LIBS, contamination of measurement windows is less critical than with other laser diagnostics because the signal intensity ratio is usually used for the elemental composition analyses. Therefore the attenuation of LIBS signals by windows is automatically cancelled. Additionally the LIBS signal intensity tends to be unaffected by the laser power in a high-laser-energy condition,

which is often called the saturation effect. Considering the measurement stability and soundness of windows, however, the cleanliness of windows has to be maintained as much as possible, especially in practical applications.

4.2 Geometric Arrangement and Measurement Species

A typical geometric arrangement of LIBS is shown in Figure 4.4. Lasers such as a pulsed Nd:YAG laser are used as a light source, and laser light is focused to the measurement point to make plasma. The emission signals from the plasma are collected and measured by a spectrometer. An image-intensified charge-coupled device (CCD) camera is mostly used to cut out unnecessary light (usually black-body-like emissions) from the hot plasma. It is worth noting that the reflection of a laser light from windows must be considered carefully. Because LIBS uses a high-energy laser light, its reflection often causes damage to optics. The reflection from plasma is sometimes tricky and malicious for LIBS systems. Plasma absorbs light and also reflects it. Damage to optics by the reflection from plasma causes troubles in some cases, especially in analyses of liquids.

Figure 4.5 shows the LIBS plasma evolution during creation and cooling processes of plasma. In the generation of plasma, the core of plasma is created by the absorption process of laser energy, such as multiphoton ionization in solids, liquids, or gases. The creation of the plasma core induces

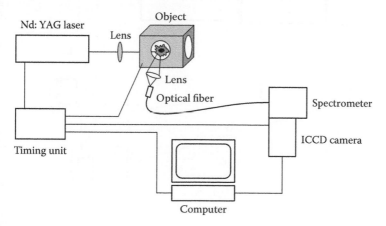

FIGURE 4.4
Typical geometric arrangement of LIBS. Main components of a LIBS system are a laser, a spectrometer, and a CCD camera. A LIBS signal is highly dependent on the plasma temperature, which means it depends on the delay time from the laser input. An ICCD camera is mostly used to select the preferable delay time for a measurement element.

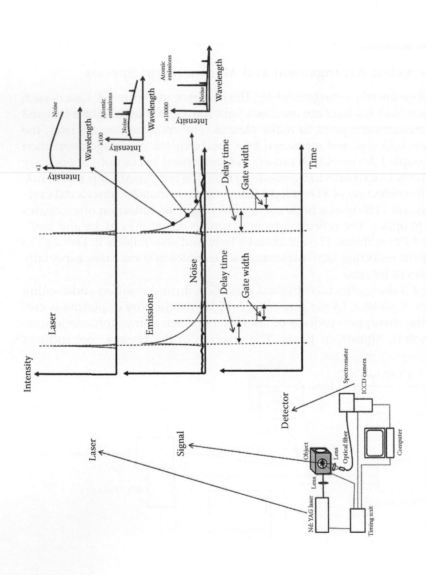

FIGURE 4.5

LIBS timing chart. After the extinction of the laser energy input, the plasma continues expanding because of its high temperature and pressure. At the same time, recombination of electrons and ions proceeds and temperature decreases gradually. LIBS signals arise in this plasma cooling period; the delay time, which corresponds to plasma temperature, is an important parameter for LIBS measurements.

the rapid growth of plasma through the absorption of the laser light by electrons in it. After the extinction of the laser energy input, the plasma continues expanding because of its high temperature and pressure. At the same time, recombination of electrons and ions proceeds and temperature decreases gradually compared to the plasma generation process. LIBS signals arise in this plasma cooling period; the delay time, which corresponds to plasma temperature, is an important parameter for LIBS measurements. The black-body-like emission, which causes the background noise, also appears depending on its plasma temperature, and there is an appropriate delay time—that is, in the plasma temperature—to get a maximum signal-to-noise ratio. Generally, the shorter delay time is chosen in emissions from higher energy levels, and the longer delay time in those from lower energy levels.

The LIBS signal intensities are illustrated in Figure 4.6. Hg is chosen as an example.[4.3] There are several emission lines just in one atom, and the identification of the emission lines is one of the most important tasks for LIBS. In practical applications, emissions are commonly attributed to more than 10 atoms. It is clear that emission intensities are much bigger at high temperature and tend to decrease according to temperature. Decrease rates depend on upper-level energies. As shown in Figure 4.6, the lower the energy, the slower the decrease. The black-body-like background emission also has temperature dependence, and it is this feature that stipulates the signal-to-noise ratio. The signal-to-noise ratio tends to become better by employing an appropriate delay time, that is, in the plasma temperature.

In applying LIBS practically, several factors shown in the previous section have to be taken into account. Figure 4.7 shows application procedures that have to be considered in LIBS applications. The most important factors for LIBS are the setting of appropriate experimental conditions and the calibration method. Although evaluation of energy levels in atoms is also an important step in choosing an appropriate emission wavelength, the main part of the procedures is dependent on experiments.

Theoretically all elements can be detected by LIBS. Sets of elements and emission lines are dependent on the application.[4.1] In the case of mineral analyses, compositions of SiO_2, CaO, MgO, Fe_2O_3, and Al_2O_3 become important factors in many applications, and these elements can be detected in the wavelength range of 240–320 nm. The emission lines of Si, Ca, Mg, Fe, and Al are summarized in Table 4.2. Sensitivity depends on the Einstein A coefficients and upper-level energies of emissions. Atoms with emissions that have both low upper-level energy and large Einstein A coefficients tend to have a good sensitivity. These atoms include Na, K, and Be, and they have much better sensitivity than other atoms. There are several ways of enhancing the detectability of LIBS. Among them are double-pulse LIBS and short-pulse LIBS methods. The double-pulse LIBS uses two lasers. The first laser pulse forms plasma and the second enhances it. The short-pulse LIBS

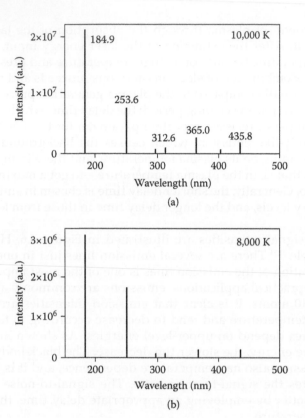

FIGURE 4.6
The LIBS signal intensities of Hg. There are several emission lines just in one atom, and the emission intensity of each line has strong temperature dependence. Choosing appropriate delay time and gate width is important to get a maximum signal-to-noise ratio. Generally, the shorter delay time is chosen in emissions from higher energy levels, and the longer delay time in those from lower energy levels. The upper energy of each emission line is 54,068 cm^{-1} at 184.9 nm, 39,412 cm^{-1} at 253.6 nm, 71,396 cm^{-1} at 312.6 nm, 71,431 cm^{-1} at 365.0 nm, and 62,350 cm^{-1} at 435.8 nm. Hg emission intensities at (a) 10,000 K; (b) 8,000 K; (c) 6,000 K; (d) 4,000 K. (e) Temperature dependence on each emission line.

method often uses femtosecond (fs) lasers, and its plasma generation process differs from that of nanosecond lasers. Efficacy of these methods depends on measurement conditions. There are also combinations of LIBS with laser-induced fluorescence (LIF) or time-of-flight mass spectrometry (TOFMS). Although these methods greatly enhance the sensitivity, their systems tend to become complicated and expensive.

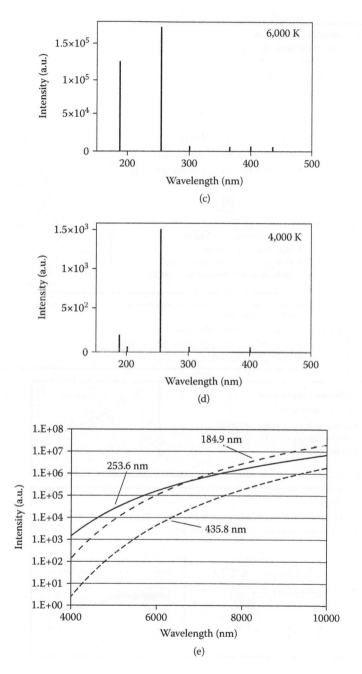

FIGURE 4.6
(Continued)

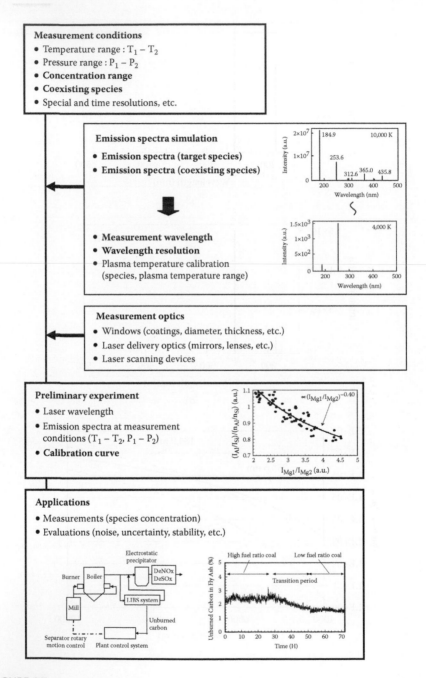

FIGURE 4.7
LIBS application procedures. The most important factors for LIBS are the settings of appropriate experimental conditions and the calibration method. Though evaluation of energy levels in atoms is also an important step in choosing an appropriate emission wavelength, the main part of the procedure is dependent on experiments.

TABLE 4.2

LIBS Emission Lines of Si, Ca, Mg, Fe, and Al

Elements	Emission Wavelength
Si	251.7
	288.1
Ca	315.9
	318.0
Mg	279.6
	285.2
Fe	262.9
	274.7
Al	309.4
	308.2

Note: The compositions of SiO_2, CaO, MgO, Fe_2O_3 and Al_2O_3 are important in many applications. The emission lines of Si, Ca, Mg, Fe, and Al appear between 260 nm and 320 nm. Emission lines of Mg (279.6 nm) and Mg (285.2 nm) can be used for plasma temperature correction.

4.3 LIBS Applications to Industrial Fields

Because LIBS is intrinsically an elemental analysis method, it has been applied to quite a few fields, including combustions, materials, toxics, foods, and so on.[4.4]–[4.14] One of the most remarkable features of LIBS is the *in situ* measurement capability. Objects to be measured are examined directly, and the pretreatment process, which becomes the time-consuming part of many conventional analyses, is not necessary in LIBS. There is also another merit of LIBS in spatial resolution. As for the solid material analyses, for example, LIBS has a good spatial resolution; two-dimensional (2-D) elemental distribution on the surface of a material can be detected by scanning the laser focus point or the material position.

There are two types of approaches for applications of LIBS. One is the *in situ* measurement of elements in a process.[4.11]–[4.16] Results are utilized to understand physical phenomena of the process. They can also be used to monitor and control a process. In this case the "object" shown in Figure 4.4 can be the part of an industrial process. There are several types of such cases, and they are summarized in Figure 4.8. LIBS can be applied to transport pipes for the process monitoring. LIBS can also be applied to materials on a moving belt conveyer, as shown in Figure 4.8(b). It is also possible to measure multipoint data by moving a measurement unit or by scanning a laser. In these applications, the vulnerability of lasers is the primary drawback to industrial applications, especially for long-term use.

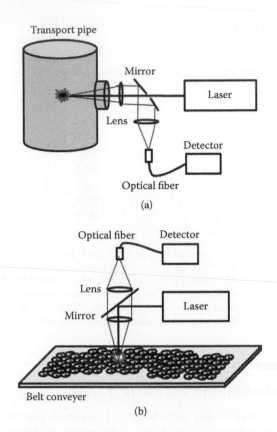

FIGURE 4.8
Several types of LIBS systems for process monitoring and control. LIBS can be used for the *in situ* measurement of elements in a process. It is also possible to measure multipoint data by moving a measurement unit or by scanning a laser. (a) LIBS system to analyze materials in transport pipes for the process monitoring. (b) LIBS system to analyze materials on a moving belt conveyer.

The other approach is the application as an elemental analyzer.[4.17]–[4.19] A sample is placed in the measurement section of a LIBS system, and elemental composition is measured instantaneously. This method has a great advantage over conventional analyses in terms of time. 2-D elemental distribution on the surface of a material has often been demonstrated in many fields. Figure 4.9 shows a typical LIBS system for 2-D elemental distribution measurements. It is possible to measure microstructures on a metal surface using an appropriate laser-scanning method. In these applications the vulnerability of lasers is less important compared to the former application.

There are several conventional methods that compete with LIBS because LIBS is intrinsically an elemental analysis. Inductively coupled plasma-atomic emission spectroscopy (ICP-AES), inductively coupled plasma-mass

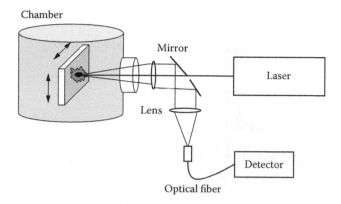

FIGURE 4.9
Microstructure measurement on metal surface using a laser-scanning method. A sample is placed in the measurement section of a LIBS system, and elemental composition is measured instantaneously. 2-D elemental distribution on the surface of a material has often been demonstrated in many fields.

spectrometry (ICP-MS), and x-ray fluorescence analysis (XRF) are typical methods that have been applied to similar fields as LIBS applications. In the real-time monitoring applications of elemental composition, XRF often competes with LIBS. The merits of LIBS are a fast response, an excellent spatial resolution, and a wide range of measurement conditions, including *in situ* measurement. It is rather rare for LIBS to compete with other laser diagnostics.

4.3.1 Engine Applications

In engine applications LIBS has been used to measure fuel-air ratios in combustion. If the fuel composition does not change, the fuel-air ratio can be inferred from the elemental analysis of unburned and burned gases. It is useful to know that the equivalence ratio can be inferred from burned gas measurement because the elemental composition does not change during reactions. LIBS can be also used for the elemental analysis of particles, such as soot, which contains not only carbon but also metallic elements. Several metallic compositions exist in soot, such as Fe, Mg, Ca, Cu, and Zn, which come from engine wear or lubricant.[4.20] It is possible to measure size-classified particles by combining LIBS with a particle classifier.[4.21] Combination of LIBS with TOFMS has also been demonstrated to measure nanoparticles,[4.22] which is shown in Chapter 7.

Figure 4.10 shows an application to measure equivalence ratio in engine exhaust.[4.23] The equivalence was monitored in the exhaust manifold of a four-cylinder engine close to the exhaust port of a cylinder. A frequency-doubled pulse Nd:YAG laser (532 nm, 100 mJ/p) was used to generate the plasma, and the LIBS signal was detected by an ICCD camera. The wavelength region

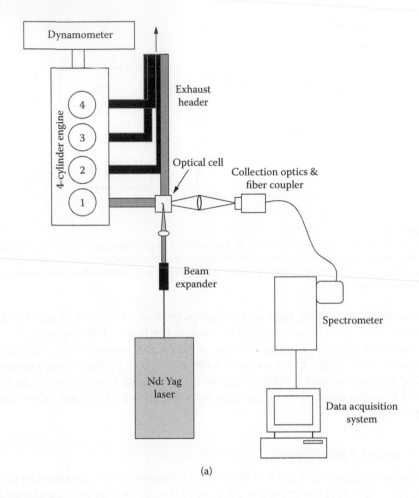

(a)

FIGURE 4.10
Application to measure equivalence ratio in engine exhaust. (a) Experimental apparatus. (b) Ratio of the integrated N intensity between 740 and 748 nm to the 711 nm C line intensity. Single-shot LIBS measurements obtained in lean and rich conditions ($W = 0.95$ and 1.26 respectively). A clear separation between the two equivalence ratios is visible. (Reprinted from F. Ferioli, S.G. Buckley, and P.V. Puzinauskas, *International Journal of Engine Research*, 7(6), 447, 2006. With permission from IMechE.)

from 700 nm to 790 nm was selected to measure atomic lines of C (711nm), N (741, 743, and 746 nm), and O (777 nm). The ratio of the C signal to N and O is related to the equivalence ratio, because the C signal originates from the fuel and the N and O signals from air. Single-shot measurement results were shown in Figure 4.10(b). Though the single-shot measurements contain some noise, the clear difference of the C to N signal ratio can be recognized between measurements obtained in lean and rich conditions. The same method can be applied to in-cylinder measurements.[4.24],[4.25]

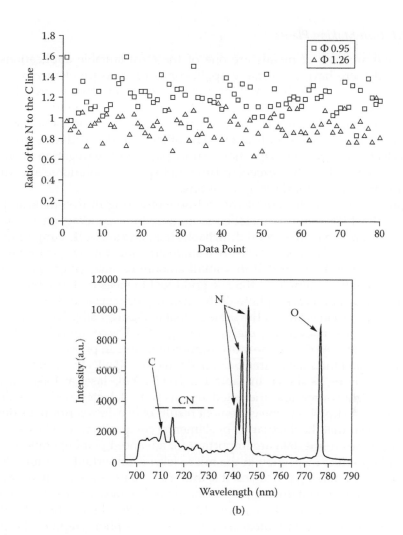

FIGURE 4.10
(Continued)

4.3.2 Plant Applications

Because of the simplicity of an apparatus, LIBS has been actively applied to commercial plants such as iron-making plants, thermal power plants, waste disposal plants, and so on. Many applications have been successfully demonstrated to monitor plant control factors using LIBS. The long-term stability and durability of LIBS devices, especially lasers, is one of the challenges in LIBS. LIBS uses pulsed lasers, and their lifetimes often limit plant applications, especially long-term continuous use for plant monitoring and control.

4.3.2.1 Iron-Making Plants

Elemental analyses of metals are one of the most suitable applications of LIBS, and there have been many applications to measure elemental compositions of iron in an iron-making process. There are two approaches to applying LIBS to iron-making plants. One is the direct monitoring of raw ores or irons to optimize the process. The *in situ* characteristics of LIBS are actively utilized in these applications. The other is the detailed measurement of products as a product inspection. An inspection of segregation is one of these examples. LIBS has excellent time and spatial resolutions, and these features are suitable for these applications.

Figure 4.11 shows an example of on-line monitoring in iron-making processes.[4.26] A LIBS unit was installed to analyze bulk minerals on a moving belt conveyor. There are several drawbacks to analyzing bulk samples using LIBS because LIBS can only measure a small amount of materials located on the surface of bulk samples. If this small amount of material on the surface can represent bulk samples, LIBS is a good tool to measure these bulk samples. Figure 4.11(a) shows a photo of the measurement system. The measurement configuration is almost the same as that of Figure 4.8(b). The phosphate rock on the belt conveyor was analyzed using LIBS. The compositions of CaO, MgO, Fe_2O_3, and Al_2O_3 are the most important factors in phosphate analysis. These emission lines have often been used in other applications such as coal and ash analyses, as shown in Table 4.2. The Nd:YAG laser at 1064 nm was used as a light source, and the gated signals were detected with 1 μs delay time. Figure 4.11(b) shows measurement results of five days continuous detection. It is clear that the fluctuation of elemental compositions can be detected by LIBS; these on-line data are important for advanced plant operations.

A rapid product inspection at the micro level is important to improve the control of steel-making processes. It is important to use measurement results for process adjustment during inspection time. The measurement concept is the same as that of Figure 4.9. Figure 4.12 shows the 2-D distribution of carbon content on a metal surface measured by LIBS.[4.17] A photograph of the crystal structure revealed by sample polishing and etching is also shown in the figure. A photo-diode pumped Nd:YLF laser at 1047 nm was used as a light source and it was operated at 1 kHz to achieve fast scans of the measured area. A 4×1 mm^2 area with a step size of 20 μm was measured in 2 min. The crystal structure photograph and the 2-D carbon map by LIBS show excellent agreement, showing the ferritic zone and the ferritic grain boundaries.

4.3.2.2 Thermal Power Plants

As for the thermal power plant applications,[4.2],[4.12],[4.27]–[4.32] coal and fly ash have been intensively investigated to apply LIBS to coal combustion applications.[4.2],[4.12],[4.27]–[4.30] Coal has been an important fuel for human beings, and it is now more and more important to make the combustion efficiency higher in order to cope with global warming. From this point of view, fly

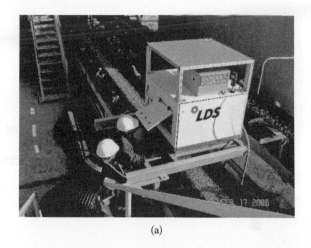

(a)

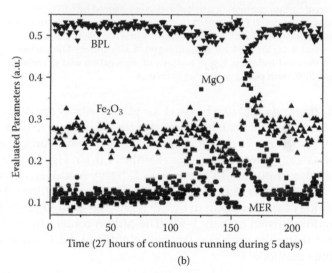

(b)

FIGURE 4.11

On-line monitoring in iron-making processes. (a) A photograph of the LIBS system on moving belt conveyer. (b) Measurement results of 5 days continuous detection. MgO (■), Fe_2O_3 (▲), BPL (▼), and MER (●) as functions of time during continuous run (total running time is 27 h) captured by the LIBS analyzer. Each analytical point results from averaging 300 laser pulses, laser energy is 40 mJ/pulse, laser frequency is 5 Hz, spectrometer settings for the delay from the laser pulse is 1 μs, and the acquisition time is 9 ms. In order to be at the same scale, the MgO concentrations were divided by 5, Fe_2O_3 concentrations by 4, and BPL concentrations by 120. BPL (bone phosphate of lime): a traditional reference to the amount (by weight percentage) of calcium phosphate contained in phosphate rock. MER = 2.185 [MER: $(Fe_2O_3 + Al_2O_3 + MgO)/$ BPL]. (Reprinted from *Spectrochimica Acta Part B*, 62(12), M. Gaft, I. Sapir-Sofer, H. Modiano, and R. Stana, "Laser induced breakdown spectroscopy for bulk minerals online analyses," 1496–1503, Copyright 2007, with permission from Elsevier.)

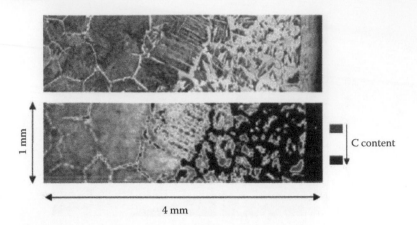

FIGURE 4.12

2-D distribution of carbon content on a metal surface measured by LIBS. Comparison between the photographs of the crystal structure revealed by sample polishing and etching (top) and the C map generated by LIBS measurement on the same sample surface (bottom). (Reprinted from *Spectrochimica Acta Part B*, 63(10), F. Boué-Bigne, "Laser-induced breakdown spectroscopy applications in the steel industry: Rapid analysis of segregation and decarburization," 1122–1129, Copyright 2008, with permission from Elsevier.)

ash analysis methods that can be used for plant controls are necessary for a better and cleaner use of coal. It is also important to reduce toxic materials in exhaust. Coal is a kind of mineral, and it contains not only carbon and hydrogen but also several heavy metals (for example, Cd, Cr, Hg, and Pb, as minor species). It is also important to monitor these heavy metals in exhaust.[4.32]

The main components of fly ash are Si, Al, Fe, Ca, and carbon. The typical analytical value of each element in fly ash is SiO_2: 60%, Al_2O_3: 20%, Fe_2O_3: 5%, CaO: 5%, and unburned carbon: 1–5%. The carbon content in the fly ash can be calculated using the emission intensities of Si, Al, Fe, Ca, and carbon by the following equation:[4.2],[4.12]

$$\text{Unburned carbon in fly ash} = \frac{\alpha_c I_c/I_{si}}{1 + \alpha_c I_c/I_{si} + \alpha_{Al} I_{Al}/I_{si} + \alpha_{Fe} I_{Fe}/I_{si} + \alpha_{Ca} I_{Ca}/I_{si}}$$

(4.2)

Here, α_i is a variable factor related to species i, and it contains the plasma temperature-correction factors. These parameters were determined under the operating conditions used in a thermal power plant.

Figure 4.13(a) shows the system for an optimal boiler control application.[4.12] The LIBS apparatus has been installed at the economizer outlet at a 1000 MW pulverized coal-fired power plant. The unburned carbon was measured for the boiler control. Measured unburned carbon values were processed and used for the optimum boiler control. Unburned carbon values were used to control the mill. The mill rotary separator can control

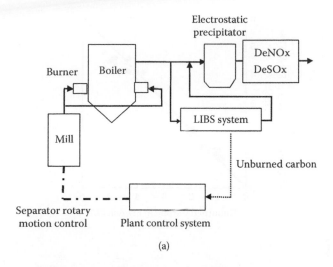

(a)

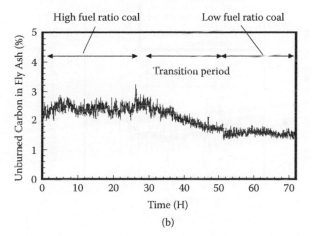

(b)

FIGURE 4.13
LIBS application to optimal boiler control. Unburned carbon in fly ash can be calculated from the major species concentrations in fly ash, which are Si, Al, Fe, Ca, and C. The transition from high to low fuel ratio of coal caused the change of unburned carbon values. The speed of the rotary separator was controlled using real-time unburned carbon measurement value. (a) System for optimal boiler control. (b) Measurement results of unburned carbon in transition from high to low fuel ratio of coals. (c) Comparison between the LIBS and the standard methods. (d) Plant control results based on unburned carbon. (Reprinted from M. Kurihara, K. Ikeda, Y. Izawa, Y. Deguchi, and H. Tarui, "Optimal boiler control through real-time monitoring of unburned carbon in fly ash by laser-induced breakdown spectroscopy," *Applied Optics*, 42(30), 6159–665, 2003. With permission of Optical Society of America.)

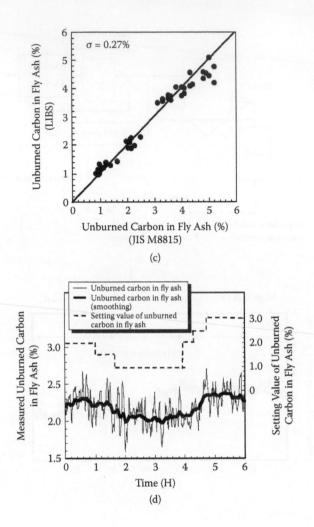

FIGURE 4.13
(Continued)

the particle size of the pulverized coal. The smaller the particle size, the lower the unburned carbon. Figure 4.13(b) presents measurement results of unburned carbon in transition from high to low fuel ratio of coal. The change of unburned carbon in fly ash resulting from the switchover can be seen in the corresponding measurement results. Over 20 types of coal were used during the operating period. Figure 4.13(c) shows a comparison between the LIBS and the standard methods. Good agreement is also shown between these measurement results. Figure 4.13(d) shows examples of plant control results based on unburned carbon. The speed of the rotary separator was raised through real-time unburned carbon control, and the value of unburned carbon was reduced, thereby increasing on-site efficiency.

Conversely, increasing the unburned carbon value reduces auxiliary power, such that operations can be performed in consideration of total plant running costs.

4.3.2.3 Waste Disposal and Recycling Plant Applications

Waste disposal and recycling plants are important facilities to achieve a sustainable society. There are several places where LIBS can be applied to these plants.[4.33]–[4.37] For waste disposal and recycling processes, wastes are inputs to the processes, and their compositions intrinsically fluctuate. Therefore it is important to know about the behavior of the waste composition to operate the system efficiently. There are also LIBS applications to monitor important elements during the disposing and recycling processes. These applications include wood recycling processes,[4.33],[4.34] plastic waste management,[4.35]–[4.37] and automobile catalyst recycling processes.[4.38] Platinum, palladium, and rhodium concentration has been measured in recycling processes of automobile catalyst scrap. Figure 4.14 shows the measurement results of chromate copper arsenate (CCA) treated and untreated woods.[4.33] CCA is a widely used preservative for woods. CCA-treated woods contain high levels of chromium, copper, and arsenate, and the separation of CCA-treated woods from normal (untreated) ones has been necessary for clean recycling of woods. LIBS can detect the elemental contents of CCA, and it is possible to distinguish CCA-treated and untreated woods using LIBS.

4.3.2.4 Other Plant Applications

LIBS has been applied in the various plants including cement plants,[4.39] nuclear facilities,[4.40],[4.41] and dairy products plants.[4.42] Figure 4.15 shows the schematics and photograph of a long-term Na and K monitoring system. In this application a LIBS unit was placed in a 20,000 ton/day cement plant to test its measurement capability under dusty commercial conditions. [4.39] The LIBS unit was placed in an exhaust duct of the plant and Na and K concentrations of the pulverized coal kiln burner exhaust gas (380 K) were monitored for a one-year period. The signals corresponding to Na, K, N, and noise signals—that is, continuous background emissions—were separated by an optical filter and detected by gated photomultipliers. Na and K signals are normalized to correct the plasma temperature effects. Na and K concentrations were also measured by the conventional method to compare with the LIBS results. Figure 4.15(c) shows the long-term Na and K monitoring results. The exhaust gas normally contained 10–50 ppb Na and K, and there appeared several sharp peaks of Na and K, corresponding to the operating conditions of the plant. The concentrations of Na and K changed similarly, because the main source of Na and K consisted of fine particles of cement material, which contained similar amounts of Na and K. LIBS was capable of detecting these sharp rises because of the 1-min

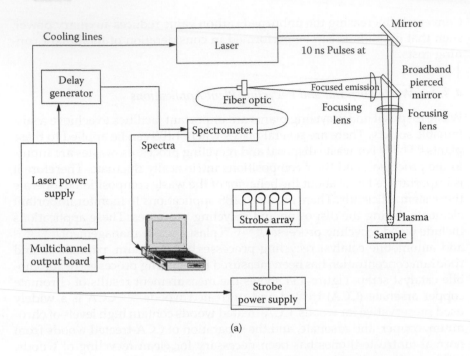

(a)

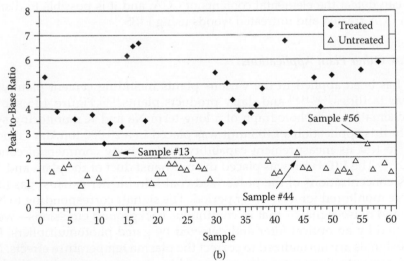

(b)

FIGURE 4.14
Measurement results of CCA-treated and untreated woods. (a) Schematic diagram for final LIBS system; (b) Measurement results of CCA-treated and untreated woods showing the ability of LIBS to detect treated and untreated wood using a peak-to-base ratio of 2.6 (10-shot average). (Reprinted from *Waste Management*, 24(4), H.M. Solo-Gabriele, T.G. Townsend, D.W. Hahn, T.M. Moskal, N. Hosein, J. Jambeck, and G. Jacobi, "Evaluation of XRF and LIBS technologies for on-line sorting of CCA-treated wood waste," 413–424, Copyright 2004, with permission from Elsevier.)

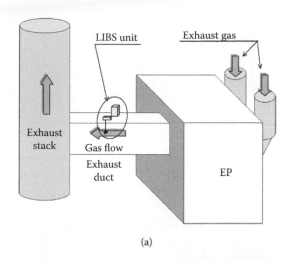

(a)

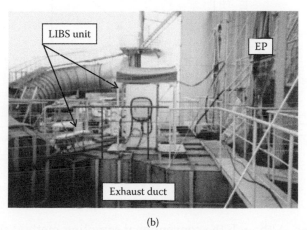

(b)

FIGURE 4.15
Schematics and measurement results of long-term Na and K monitoring. 300 mJ/p YAG laser output (1064 nm) was used to detect Na and K. Emission of 589 nm was used for Na, and 767 nm for K. 100 μs delay time for Na and 120 μs for K were selected to reduce the background plasma emission noise. LIBS is capable of detecting ppb-level concentration change due to its 1-min detection time. (a) Schematics of a long-term Na and K monitoring system; (b) Photograph of a long-term Na and K monitoring system; (c) Measurement results for Na; (d) Comparison between LIBS and the conventional Method for Na. (Reprinted from Y. Deguchi, M. Noda, Y. Fukuda, Y. Ichinose, Y. Endo, M. Inada, Y. Abe, and S. Iwasaki, *Measurement Science and Technology*, 13(10), R103, 2002. With permission.)

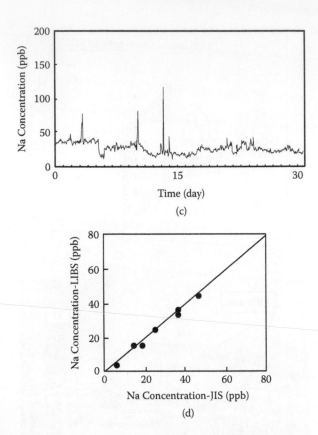

FIGURE 4.15
(Continued)

detection time. Although the conventional sampling method requires several hours to monitor a few ppb of Na, LIBS can detect the same concentration within a 1-min measurement time and exhibits excellent monitoring capability.

4.3.3 Other Applications

Applications of LIBS have occurred in many industrial fields, including analyses from food to nuclear materials. Since LIBS is a method to analyze elemental compositions, industrial fields with a need for elemental analysis are potential LIBS applications.[4.43] As for LIBS application to foods and water,[4.44]–[4.46] measurements of compositions or contaminations have been demonstrated in flours, wheat, barley, and water. Soils[4.47] and minerals[4.48] are among applications of LIBS. Many elements such as Mg, Al, Cu, Cr, K, Mn, Rb, Cd, and Pb have been measured by LIBS. LIBS has been also applied to the safety and security fields,[4.49],[4.50] including explosives.[4.50]

4.4 Future Developments

There are many other LIBS applications in various fields. LIBS is intrinsically an elemental analysis method, and it can be applied to industrial fields with a need for elemental analyses. One of the major drawbacks of LIBS is the difficulty of quantitative analysis. There are numerous correction methods for LIBS to achieve quantitative information; however, they are usually application dependent and there is no universal method applicable to all LIBS applications because of the complexities of plasma characteristics produced by the LIBS process. With this in mind, databases for LIBS in terms of quantitative analyses, which can be usable by not only spectroscopists but designers for various fields, will be important for the advancement of industrial applications. With these databases, engineers of various application fields can efficiently employ the LIBS technique. It is also important to understand that LIBS can be used for the molecular detection employing the "elemental fingerprint of a molecule." It often happens that the elemental signal pattern of some molecules is unique, and this pattern can be used to evaluate the target molecule. In these applications, such a systematic database is valuable.

Many of the LIBS applications have used pulsed lasers and ICCD detectors. Because of their high cost and vulnerability, it is rather difficult to use those laser systems for the control of industrial processes for long-term use. Lasers, especially, are not rugged enough to use them continuously for several years, which is often required in several industries. Therefore the advancement of lasers and detectors is crucial for these applications. Currently, portable LIBS systems have been developed and applied to various fields. Portability of the system is important for many applications.[4.49],[4.50] The portable system is useful for large-area measurements. For example, monitoring of radioactive elements[4.51] in large areas becomes important in case of leakage of radioactive materials from a nuclear power plant. This has actually happened in the Japanese nuclear power plants in Fukushima in 2011. LIBS has the potential to cover these applications.

References

[4.1]   A.W. Miziolek, V. Palleschi, and I. Schechter, *Laser Induced Breakdown Spectroscopy*, Cambridge, Cambridge University Press, 2008.

[4.2]   M. Noda, Y. Deguchi, S. Iwasaki, and N. Yoshikawa, "Detection of carbon content in a high-temperature and high-pressure environment using laser-induced breakdown spectroscopy," *Spectrochimica Acta B*, 57(4), 701–709, 2002.

[4.3]   R. Payling and P. Laerkins, *Optical Emission Lines of the Elements*, New York, John Wiley & Sons, 2000.

[4.4]   L.W. Peng, W.L. Flower, K.R. Hencken, H.A. Johnsen, R.F. Renzi, and N.B. French, "A laser-based technique for continuously monitoring metal emissions from thermal waste treatment units," *Process Control and Quality*, 7(1), 39–49, 1995.

[4.5]   S. Yalcin, D.R. Crosley, G.P. Smith, and G.W. Faris, "Spectroscopic characterization of laser-produced plasmas for in situ toxic metal monitoring," *Hazardous Waste and Hazardous Materials*, 13(1), 51–61, 1996.

[4.6]   A.E. Pichahchy, D.A. Cremers, and M.J. Ferris, "Elemental analysis of metals under water using laser-induced breakdown spectroscopy," *Spectrochimica Acta Part B*, 52(1), 25–39, 1997.

[4.7]   L. St-Onge, M. Sabsabi, and P. Cielo, "Analysis of solids using laser-induced plasma spectroscopy in double-pulse mode," *Spectrochimica Acta Part B*, 53(3), 407–415, 1998.

[4.8]   R.E. Neuhauser, P. Panne, and R. Niessner, "Laser-induced plasma spectroscopy (LIPS): A versatile tool for monitoring heavy metal aerosols," *Analytica Chimica Acta*, 392(1), 47–54, 1999.

[4.9]   C. Haisch, R. Niessner, O.I. Matveev, U. Panne, and N. Omenetto, "Element-specific determination of chlorine in gases by laser-induced-breakdown-spectroscopy (LIBS)," *Fresenius' Journal of Analytical Chemistry*, 356, 21–26, 1996.

[4.10]   R.E. Neuhauser, P. Panne, R. Niessner, G.A. Petrucci, P. Cavalli, and N. Omenetto, "Online and in-situ detection of lead aerosols by plasma-spectroscopy and laser-excited atomic fluorescence spectroscopy," *Analytica Chimica Acta*, 346(1), 37–48, 1997.

[4.11]   U. Panne, C. Haisch, M. Clara, and R. Niessner, "Analysis of glass and glass melts during the vitrification process of fly and bottom ashes by laser-induced plasma spectroscopy. Part I: Normalization and plasma diagnostics," *Spectrochimica Acta Part B*, 53(14), 1957–1968, 1998.

[4.12]   M. Kurihara, K. Ikeda, Y. Izawa, Y. Deguchi, and H. Tarui, "Optimal boiler control through real-time monitoring of unburned carbon in fly ash by laser-induced breakdown spectroscopy," *Applied Optics*, 42(30), 6159–665, 2003.

[4.13]   M. Gaft., E. Dvir, H. Modiano, and U. Schone, "Laser induced breakdown spectroscopy machine for online ash analyses in coal," *Spectrochimica Acta Part B*, 63, 1177, 2008.

[4.14]   M. Pouzar, T. Cernohorsky, M. Prusova, P. Prokopcakov, and A. Krejcova, "LIBS analysis of crop plants," *Journal of Analytical Atomic Spectrometry*, 24(7), 953–957, 2009.

[4.15]   M.A. Gondal, T. Hussain, Z.H. Yamani, and M.A. Baig, "On-line monitoring of remediation process of chromium polluted soil using LIBS," *Journal of Hazardous Materials*, 163(2–3), 1265–1271, 2009.

[4.16]   O.T. Butler, W.R.L. Cairns, J.M. Cook, and C.M. Davidson, "Atomic spectrometry update. Environmental analysis," *Journal of Analytical Atomic Spectrometry*, 25(2), 103–141, 2010.

[4.17]   F. Boué-Bigne, "Laser-induced breakdown spectroscopy applications in the steel industry: Rapid analysis of segregation and decarburization," *Spectrochimica Acta Part B*, 63(10), 1122–1129, 2008.

[4.18]   M. Galiová, J. Kaiser, K. Novotný, J. Novotný, T. Vaculovic, M. Liška, R. Malina, K. Stejskal, V. Adam, and R. Kizek, "Investigation of heavy-metal accumulation in selected plant samples using laser induced breakdown spectroscopy and laser ablation inductively coupled plasma mass spectrometry," *Applied Physics A*, 93(4), 917–922, 2008.

[4.19] O. Samek, D.C.S. Beddows, H.H. Telle, J. Kaiser, M. Liska, J.O. Caceres, A. G. Urena, "Quantitative laser-induced breakdown spectroscopy analysis of calcified tissue samples," *Spectrochimica Acta Part B*, 56(6), 865–875, 2001.

[4.20] K. Lombaert, S. Morel, L. Le Moyne, P. Adam, J. Tardieu de Maleissye, and J. Amouroux, "Nondestructive analysis of metallic elements in diesel soot collected on filter: Benefits of laser induced breakdown spectroscopy," *Plasma Chemistry and Plasma Processing*, 24(1), 41–56, 2004.

[4.21] N. Strauss, C. Fricke-Begemann, and R. Noll, "Size-resolved analysis of fine and ultrafine particulate matter by laser-induced breakdown spectroscopy," *Journal of Analytical Atomic Spectroscopy*, 25(6), 867–874, 2010.

[4.22] Y. Deguchi, N. Tanaka, M. Tsuzaki, A. Fushimi, S. Kobayashi, and K. Tanabe, "Detection of components in nanoparticles by resonant ionization and laser breakdown time-of flight mass spectroscopy," *Environmental Chemistry*, 5(6), 402–412, 2008.

[4.23] F. Ferioli, S.G. Buckley, and P.V. Puzinauskas, "Real-time measurement of equivalence ratio using laser-induced breakdown spectroscopy," *International Journal of Engine Research*, 7(6), 447–457, 2006.

[4.24] F. Ferioli, P.V. Puzinauskas, and S.G. Buckley, "Laser-induced breakdown spectroscopy for on-line engine equivalence ratio measurements," *Applied Spectroscopy*, 57(9), 1183–1189, 2003.

[4.25] S. Joshi, D.B. Olsen, C. Dumitrescu, P.V. Puzinauskas, and A.P. Yalin, "Laser-induced breakdown spectroscopy for in-cylinder equivalence ratio measurements in laser-ignited natural gas engines," *Applied Spectroscopy*, 63(5), 549–554, 2009.

[4.26] M. Gaft, I. Sapir-Sofer, H. Modiano, and R. Stana, "Laser induced breakdown spectroscopy for bulk minerals online analyses," *Spectrochimica Acta Part B*, 62(12), 1496–1503, 2007.

[4.27] T. Ctvrtnickova, M.-P. Mateo, A. Yañez, and G. Nicolas, "Characterization of coal fly ash components by laser-induced breakdown spectroscopy," *Spectrochimica Acta Part B*, 64(10), 1093–1097, 2009.

[4.28] M.P. Mateo, G. Nicolas, and A. Yañez, "Characterization of inorganic species in coal by laser-induced breakdown spectroscopy using UV and IR radiations," *Applied Surface Science*, 254(4), 868–872, 2007.

[4.29] T. Ctvrtnickova, M.P. Mateo, A. Yañez, and G. Nicolas, "Laser induced breakdown spectroscopy application for ash characterisation for a coal fired power plant," *Spectrochimica Acta Part B*, 65(8), 734–737, 2010.

[4.30] M. Gaft, E. Dvir, H. Modiano, and U. Schone, "Laser induced breakdown spectroscopy machine for online ash analyses in coal," *Spectrochimica Acta Part B*, 63(10), 1177–1182, 2008.

[4.31] L.G. Blevins, C.R. Shaddix, S.M. Sickafoose, and P.M. Walsh, "Laser-induced breakdown spectroscopy at high temperatures in industrial boilers and furnaces," *Applied Optics*, 42(3), 6107–18, 2003.

[4.32] D.L. Monts, J.P. Singh, Y.S. Abhilasha, H. Zhang, F.Y. Yueh, P.R. Jang, and S.K. Singh, "Toward development of a laser-based continuous emission monitor system for toxic metals in off-gases," *Combustion Science and Technology*, 134(1–6), 103–126, 1998.

[4.33]   H.M. Solo-Gabriele, T.G. Townsend, D.W. Hahn, T.M. Moskal, N. Hosein, J. Jambeck, and G. Jacobi, "Evaluation of XRF and LIBS technologies for on-line sorting of CCA-treated wood waste," *Waste Management*, 24(4), 413–424, 2004.

[4.34]   T.M. Moskal and D.W. Hahn, "On-line sorting of wood treated with chromated copper arsenate using laser-induced breakdown spectroscopy," *Applied Spectroscopy*, 56(10), 1337–1344, 2002.

[4.35]   M.N. Siddiqui, M.A. Gondal, and M.M. Nasr, "Determination of trace metals using laser induced breakdown spectroscopy in insoluble organic materials obtained from pyrolysis of plastics waste," *Bulletin of Environmental Contamination and Toxicology*, 83, 141–145, 2009.

[4.36]   J. Anzano, B. Bonilla, B. Montull-Ibor, R.-J. Lasheras, and J. Casas-Gonzalez, "Classifications of plastic polymers based on spectral data analysis with laser induced breakdown spectroscopy," *Journal of Polymer Engineering*, 30(3–4), 177–187, 2010.

[4.37]   M.A. Gondal and M.N. Siddiqui, "Identification of different kinds of plastics using laser-induced breakdown spectroscopy for waste management," *Journal of Environmental Science and Health Part A*, 42(13), 1989–1997, 2007.

[4.38]   G. Asimellis, N. Michos, I. Fasaki, and M. Kompitsas, "Platinum group metals bulk analysis in automobile catalyst recycling material by laser-induced breakdown spectroscopy," *Spectrochimica Acta Part B*, 63(11), 1338–1343, 2008.

[4.39]   Y. Deguchi, M. Noda, Y. Fukuda, Y. Ichinose, Y. Endo, M. Inada, Y. Abe, and S. Iwasaki, "Industrial applications of temperature and species concentration monitoring using laser diagnostics," *Measurement Science and Technology*, 13(10), R103–R115, 2002.

[4.40]   M.M. Tripathi, K.E. Eseller, F.-Y. Yueh, and J.P. Singh, "Multivariate calibration of spectra obtained by laser induced breakdown spectroscopy of plutonium oxide surrogate residues," *Spectrochimica Acta Part B*, 64 (11–12), 1212–1218, 2009.

[4.41]   A. Sarkar, V.M. Telmore, D. Alamelu, and S.K. Aggarwal, "Laser induced breakdown spectroscopic quantification of platinum group metals in simulated high level nuclear waste," *Journal of Analytical Atomic Spectrometry*, 24(11), 1545–1550, 2009.

[4.42]   T. Hussain and M.A. Gondal, "Detection of toxic metals in waste water from dairy products plant using laser induced breakdown spectroscopy," *Bulletin of Environmental Contamination and Toxicology*, 80(6), 561–565, 2008.

[4.43]   R. Noll, V. Sturm, Ü. Aydin, D. Eilers, C. Gehlen, M. Höhne, A. Lamott, J. Makowe, and J. Vrenegor, "Laser-induced breakdown spectroscopy—From research to industry, new frontiers for process control," *Spectrochimica Acta Part B*, 63(10), 1159–1166, 2008.

[4.44]   F. Zhao, Z. Chen, F. Zhang, R. Li, and J. Zhou, "Ultra-sensitive detection of heavy metal ions in tap water by laser-induced breakdown spectroscopy with the assistance of electrical-deposition," *Analytical Methods*, 2, 408–414, 2010.

[4.45]   M. Pouzar, T. Cernohorsky, M. Prusova, P. Prokopcakova, and A. Krejcova, "LIBS analysis of crop plants," *Journal of Analytical Atomic Spectrometry*, 24(7), 953–957, 2009.

[4.46]   H.-H. Cho, Y.-J. Kim, Y.-S. Jo, K. Kitagawa, N. Arai, and Y.-I. Lee, "Application of laser-induced breakdown spectrometry for direct determination of trace elements in starch-based flours," *Journal of Analytical Atomic Spectrometry*, 16(6), 622–627, 2001.

[4.47]   M.A. Gondal, T. Hussain, Z.H. Yamani, and M.A. Baig, "On-line monitoring of remediation process of chromium polluted soil using LIBS," *Journal of Hazardous Materials*, 163, 1265–1271, 2009.

[4.48]   R.S. Harmon, F.C. DeLucia Jr., A. LaPointe, R.J. Winkel Jr., and A.W. Miziolek, "LIBS for landmine detection and discrimination," *Analytical and Bioanalytical Chemistry*, 385(6), 1140–1148, 2006.

[4.49]   C.A. Munson, J.L. Gottfried, E. Gibb Snyder, F.C. De Lucia Jr., B. Gullett, and A.W. Miziolek, "Detection of indoor biological hazards using the man-portable laser induced breakdown spectrometer," *Applied Optics*, 47(31), G48–G57, 2008.

[4.50]   J.L. Gottfried, F.C. De Lucia Jr., C.A. Munson, and A.W. Miziolek, "Laser-induced breakdown spectroscopy for detection of explosives residues: A review of recent advances, challenges, and future prospects," *Analytical and Bioanalytical Chemistry*, 395(2), 283–300, 2009.

[4.51]   F.R. Doucet, G. Lithgow, R. Kosierb, P. Bouchard, and M. Sabsabi, "Determination of isotope ratios using laser-induced breakdown spectroscopy in ambient air at atmospheric pressure for nuclear forensics," *Journal of Analytical Atomic Spectrometry*, 26(3), 536–541, 2011.

5

Spontaneous Raman
Spectroscopy and CARS

5.1 Principle

5.1.1 Spontaneous Raman Spectroscopy

When light is irradiated to molecules or small particles $(d \ll \lambda)$, there appear two types of scatterings: elastic scattering and inelastic scattering. The elastic scattering is called Rayleigh scattering and the inelastic one Raman scattering.[5.1] As shown in Figure 5.1, there is no energy exchange between the incident light and molecules in Rayleigh scattering; but energy exchanges corresponding to molecular rotational and/or vibrational (ro-vibrational) energies occur in Raman scattering. Because of those energy exchanges, wavelengths of Raman scattering light are shifted by those energies. The shift in energy gives information about molecules, and their spectra are often called "molecular fingerprints." The wavelength shifts are unique for each individual molecule, and thus multiple species detection is possible in many applications.

In classical theory, Raman scattering occurs based on oscillating dipole moment p induced by the incident light

$$p = \alpha E \tag{5.1}$$

where α is the polarizability and E the electric field. The image of the oscillating dipole moment is shown in Figure 5.2. The dipole moment is induced in a molecular system by the electric field of incident light. It is a common mathematical method that the polarizability is expanded as the sum of a static term α and a derivative by the coordinate Q, which means its vibrational mode

$$\alpha = \alpha_0 + \left(\frac{\partial \alpha}{\partial Q} \right)_0 Q \tag{5.2}$$

As the coordinate Q represents the vibrational mode, Q is given by

$$Q = Q_0 \cos \omega_v t \tag{5.3}$$

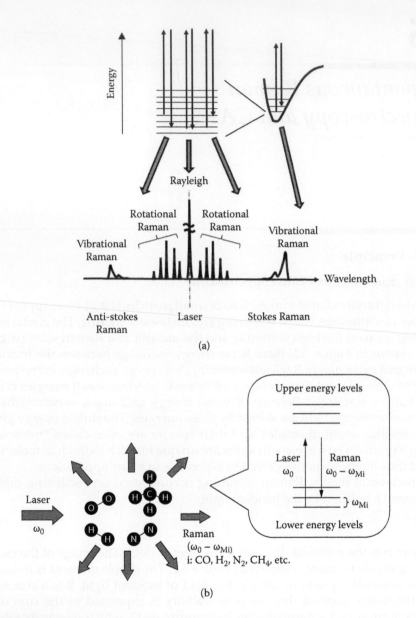

FIGURE 5.1

Elastic and inelastic scattering processes. The wavelength of Raman scattering light is shifted from that of the incident light by energies corresponding to molecular vibrational and/or rotational energies. The wavelength shifts are unique for individual molecules, and multiple species detection is possible in many applications. (a) Elastic and inelastic scattering; (b) Raman scattering.

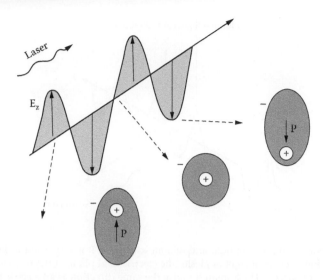

FIGURE 5.2
Image of the oscillating dipole moment in Raman process. The dipole moment is induced in a molecular system by the electric field of incident light. The dipole moment arises from the displacement of the electron cloud from the nucleus by an electric field of light.

where ω_v is the vibrational frequency of the Raman active molecule. The electric field also oscillates with the frequency of the incident light as follows:

$$E = E_0 \cos \omega_0 t \tag{5.4}$$

Using Equations (5.2) through (5.4), Equation (5.1) becomes

$$p = \left[\alpha_0 + \left(\frac{\partial \alpha}{\partial Q} \right)_0 Q_0 \cos \omega_v t \right] E_0 \cos \omega_0 t$$

$$= \alpha_0 E_0 \cos \omega_0 t + \left(\frac{\partial \alpha}{\partial Q} \right)_0 \frac{Q_0 E_0}{2} [\cos(\omega_0 - \omega_v)t + \cos(\omega_0 + \omega_v)t] \tag{5.5}$$

The first term, $\alpha_0 E_0 \cos \omega_0 t$, leads to a scattering at the incident light frequency, which means Rayleigh scattering. The second term, containing $\cos(\omega_0 - \omega_v)t$ and $\cos(\omega_0 + \omega_v)t$, indicates the scattering with the frequency shifted by ω_v, which is Raman scattering.

It is important to know that the Raman scattering appears at right angles to the electric field plane of the incident light if the incident light is polarized, as shown in Figure 5.3. It is due to this polarization characteristic that researchers sometimes have difficulty getting Raman signals. The Raman signal at the same direction as an electric field plane of the incident light—that is, at the z direction in Figure 5.3—is very weak compared to that at the x direction.

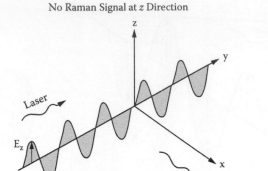

No Raman Signal at z Direction

FIGURE 5.3
Geometry of polarized incident light and Raman scattering. It is important to know that the Raman scattering appears at right angles to the electric field plane of the incident light if the incident light is polarized. The Raman signal at the same direction as an electric field plane of the incident light—that is, the z direction—is very weak compared to that at the x direction.

According to Equation (5.5), the Raman signal intensity I_{Raman} is given by

$$I_{Raman} = nI_0\left(\frac{d\sigma}{d\Omega}\right)\Omega\varepsilon V/4\pi \tag{5.6}$$

where n is the number density of Raman-active molecule, I_0 the incident light intensity, $(d\sigma/d\Omega)$ the differential scattering cross section, Ω the solid angle, ε the light collecting efficiency, and V the measurement volume.

There are ro-vibrational energy levels in molecules and these energy structures have to be included in Equation (5.6). The number density of molecule at a (v, J) ro-vibrational state is given by the Boltzmann equation

$$n_{v,j} = n\frac{g_{v,j}e^{-E_{v,j}/kT}}{\sum\limits_{v,j} g_{v,j}e^{-E_{v,j}/kT}} \tag{5.7}$$

where $g_{v,J}$ and $E_{v,J}$ are the degeneracy and the energy of (v,J) state, respectively. Thinking of the Raman scattering with $(v,J) \to (v',J')$ transitions, Equation (5.6) becomes

$$I_{Raman} = \sum_v \sum_j \left[n_{v,j} I_0 \left(\frac{d\sigma}{d\Omega}\right)^{v'j'}_{v,j} \Omega\varepsilon V/4\pi \right] \tag{5.8}$$

It is also necessary to include line broadening factors into Equation (5.8) to form Raman spectra. There are three types of line broadenings as described in Appendix D. Using the line shape function $G(\nu)$, the Raman spectra can be calculated using the following equation:

$$I_{Raman}(\nu) = \sum_{v}\sum_{j}\left[n_{v,j}I_0 \left(\frac{d\sigma}{d\Omega}\right)_{v,j}^{v'j'} G_{(v,j),(v'j')}(\nu)\Omega\epsilon V/4\pi \right] \qquad (5.9)$$

In many cases the slit function by a detection device such as a spectrometer is much broader than Doppler and collision broadenings in a Raman detection system. In this case, $G(\nu)$ is neglected and the Raman spectra can be simulated by the similar form as Equation (5.9)

$$I_{Raman}(\nu) = \sum_{v}\sum_{j}\left[n_{v,j}I_0 \left(\frac{d\sigma}{d\Omega}\right)_{v,j}^{v'j'} G_S(v_0,\nu)\Omega\epsilon V/4\pi \right] \qquad (5.10)$$

where $G_S(v_0,\nu)$ is the slit function of the detection system. The slit function often has a Gaussian line shape and has to be determined using a single-wavelength light source including a low-pressure mercury lamp.

For the evaluation of Raman spectra, selection rules can be applied to the $(v,J) \to (v',J')$ transitions. In diatomic molecules such as N_2 and O_2, it is often enough to employ the selection rules $(v \to v+1)$ and $(J \to J)$ for vibrational Raman spectra. In this case, the frequency of vibrational Raman spectra can be given by the energy conservation

$$\nu_{(v,J)\to(v+1,J)} = \nu_0 - (E_{v+1,J} - E_{v,J})/hc$$

$$= \nu_0 - (\omega_e - 2\omega_e x_e(v+1) - \alpha_e J(J+1)) \qquad (5.11)$$

where ω_e and $\omega_e x_e$ are the vibrational constants and α_e is the rotational constant (see Appendix B). The differential Raman cross sections $(d\sigma/d\Omega)$ for the transition $(v,J) \to (v+1,J)$ have the following relation:

$$\left(\frac{d\sigma}{d\Omega}\right)_{v,J}^{v+1,J} \propto (v+1)(v_0 - v_R)^4 \qquad (5.12)$$

Since $v_0 - v_R$ is the Raman scattering frequency, the Raman scattering intensity is inversely proportional to the fourth power of its wavelength. Using Equations (5.6), (5.7), and (5.9), the Raman spectra can be calculated using the molecular constants.

If all the molecular constants of Raman spectra described in Equation (5.6) through (5.11) are available, the interpretation of its signal is simple, and this makes spontaneous Raman spectroscopy an attractive method for industrial applications. On the other hand, intensities of spontaneous Raman scattering are very weak, and this limits its applicability to practical applications. Especially fluorescence interferences often make this method ineffective in many applications. Raman spectra of N_2,[5.3] H_2,[5.3] CO_2,[5.3],[5.4] and H_2O[5.3],[5.5] are shown in Figure 5.4. The shapes of their Raman spectra have

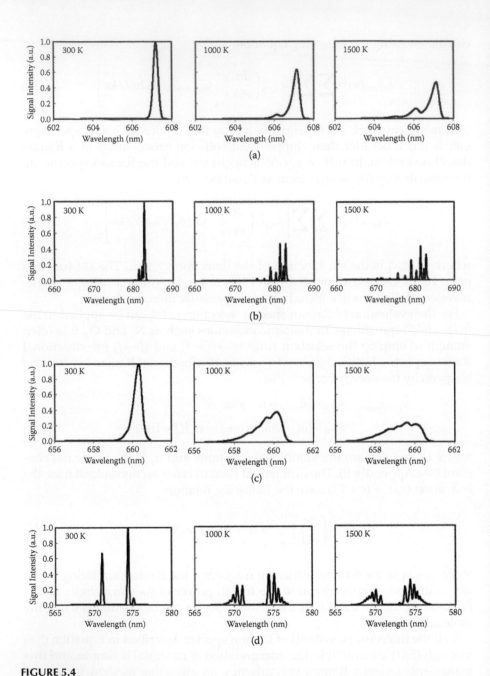

FIGURE 5.4
Raman spectra of (a) N_2, (b) H_2, (c) CO_2, and (d) H_2O. The shapes of the Raman spectra have unique structures depending on their molecular structures. Therefore multispecies detection is possible even in high-temperature and high-pressure conditions.

TABLE 5.1

Characteristics of Spontaneous Raman Spectroscopy

	Characteristics	Countermeasure
Theoretical treatments	• Energy transition	
	• Selection rule	
Temperature	• Boltzmann distribution (lower energy states)	• Theoretical evaluation
Pressure		
Windows	• Sensitive	• Purge gas
		• Heating
Calibration	• Easy	
Noise	• Fluorescence	• Incident light wavelength
	• Incident light	
Measurement item	• Temperature	
	• Concentration	
Measurement dimension	• Point (1-D)	
Detection limit	• %	• Nonlinear process (SERS)
stability	• Laser	• diode laser

Note: This method is theoretically simple and easy to handle. It is important to reduce noise as much as possible because the intensity of Raman spectra is very weak. The noise effects tend to increase with temperature.

unique structures depending on their molecular structures. Therefore the multispecies detection can be possible even in high-temperature and high-pressure conditions.

There are several key factors in obtaining Raman signals. Table 5.1 summarizes the characteristics of spontaneous Raman spectroscopy. This method is theoretically simple and easy to handle. It is important to reduce noise as much as possible because the intensity of Raman spectra is very weak. The noise effects tend to increase with temperature, and careful consideration is necessary to estimate or eliminate these parameters.

1. *Background noise.* In Raman spectroscopy, noise reduction is indispensable because of its weak signal intensity. Fluorescence interference is a nuisance phenomenon. Raman signals have a polarization property as shown in Figure 5.3, and it is important to take this into account in experimental setups. There are several techniques for the improvement of the signal-to-noise ratio. One of them is the selection of an incident light wavelength. As can be seen from Equation (5.6), Raman signals can arise using any incident light wavelengths. However, fluorescence intensities usually have strong wavelength dependence, which shows the tendency of less fluorescence noise with a longer wavelength light. Use of the incident light with longer wavelengths works well for the reduction of

fluorescence noise interference, but it is important to remember that Raman signal intensity also varies with one-fourth power of the wavelength—the incident light with longer wavelength means less Raman signal intensity. Another technique for the improvement of the signal-to-noise ratio is the time-resolved detection of Raman signals. Just as the name suggests, Raman signals are generated by the scattering phenomena, which has no time delay from the incident light. Fluorescence signals have a finite length of delay time, however, because they occur following an absorption phenomenon (see Chapter 3). Therefore, the time-gated detection method is sometimes a feasible way to reduce noise in Raman signals.

2. *Dirt on measurement windows.* As is often the case with other laser diagnostics, cleanliness of measurement windows has to be maintained. Dirt on measurement windows causes attenuation of the incident light and Raman signal intensities, increase of noise by scattering of the incident light, and damage to windows. Sometimes fluorescence from the dirt on windows causes noise. Measurement windows are necessary in many industrial applications and they require careful handling, especially in practical applications.

3. *Pressure effects.* As shown in Equation (5.6), Raman signal intensity directly depends on the number density of Raman-active molecules. Therefore Raman signal intensity increases with pressure. There exists a pressure broadening effect of each Raman scattering line; however, this effect has little consequence on the measurement. From this standpoint, the main effect of pressure on Raman signals is an increase of its signal intensity. Raman spectroscopy is one of few methods that work well in high-pressure conditions.

4. *Temperature effect and changes of population fraction in energy levels.* Population fraction in each molecular energy level is dependent on temperature based on the Boltzmann equation. The population fraction distribution in energy levels causes changes in the shape and intensity of Raman spectra as shown in Figure 5.4. In quantitative concentration and temperature measurements, it is necessary to predict these tendencies using simulation of Raman spectra in the temperature range considered.

5.1.2 CARS

Coherent anti-Stokes Raman spectroscopy (CARS) is one of the nonlinear optical processes using Raman effects.[5.1],[5.2],[5.6] The principle behind CARS is illustrated in Figure 5.5. CARS typically uses a pump beam ω_1 and a probe beam ω_2, and the blue-shifted (anti-Stokes) signal $\omega_3 = 2\omega_1 - \omega_2$ is generated by the nonlinear optical process. One of the merits of CARS is its strong signal intensity compared with that of spontaneous Raman

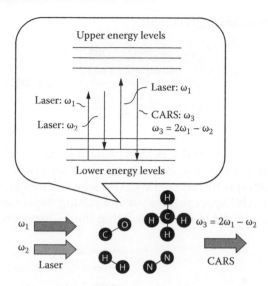

FIGURE 5.5
Principle behind CARS. CARS is one of the nonlinear optical processes. Using a pump beam ω_1 and a probe beam ω_2, the CARS signal with $\omega_3 = 2\omega_1 - \omega_2$ is generated by Raman effects.

spectroscopy. Different from other laser diagnostics, the CARS process forms a beamlike CARS signal, and this also makes CARS a robust method against noises like fluorescence. Both spontaneous Raman spectroscopy and CARS can be used for temperature and species concentration measurements.

Raman signals arise directly from the oscillating dipole moment. CARS signals are induced by the third-order susceptibility, which is a complex number. The CARS signal intensity I_{CARS} is given by

$$I_{CARS}(\omega_3 = 2\omega_1 - \omega_2) = K \, |\chi_{CARS}|^2 \, I_1^2(\omega_1) I_2(\omega_2) \qquad (5.13)$$

where K is the proportional constant, χ_{CARS} the CARS susceptibility, I_1 the pump beam intensity, I_2 the probe beam intensity. Since χ_{CARS} is proportional to the number density of Raman-active molecule n, the CARS signal intensity typically depends on n^2. χ_{CARS} is a complex number and it is given by the following form:

$$\chi_{CARS} = \chi_r + \chi_{nr}$$

$$= \sum_j \left(\chi_j' + i\chi_j'' \right) + \chi_{nr} \qquad (5.14)$$

where χ_r and χ_{nr} are the resonant and nonresonant susceptibilities, respectively. χ' and χ'' are real and imaginary parts of the resonant susceptibility and they have the following relations:

$$\chi'_j \propto \frac{2}{4} \frac{\omega_j \Gamma_j}{\omega_j^2 + \Gamma_j^2} \left(\frac{d\sigma}{d\Omega}\right)_j \tag{5.15}$$

$$\chi''_j \propto \frac{\Gamma_j^2}{4} \frac{1}{\omega_j^2 + \Gamma_j^2} \left(\frac{d\sigma}{d\Omega}\right)_j \tag{5.16}$$

where $\Delta\omega_j = \omega_j - (\omega_1 - \omega_2)$ and Γ is the constant related to the spontaneous Raman line width through $\Gamma = 2\pi\Delta\nu_R$. It is important to know the roles of χ', χ'', and χ_{nr} in CARS spectra. Since the CARS signal intensity is proportional to $|\chi_{CARS}|^2$, the CARS spectra have an interesting aspect as illustrated in Figure 5.6. It is important to understand that the χ_{nr} plays an important role

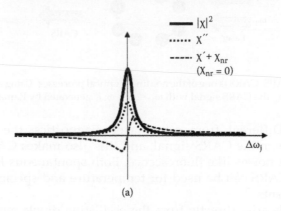

(a)

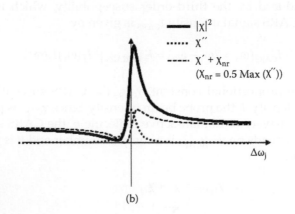

(b)

FIGURE 5.6
Influence of χ_{nr} on CARS spectra. It is important to understand that the χ_{nr} plays an important role in CARS spectra, and the shape of CARS spectra changes according to the influence of χ_{nr}. When the concentration of the measured species becomes low, this effect is significantly enhanced and often limits its sensitivity. (a) $\chi_{nr} = 0$; (b) $\chi_{nr} = 0.5$ Max(χ'').

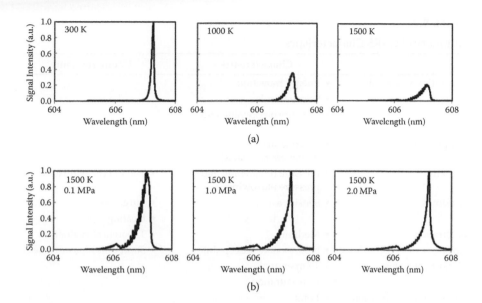

FIGURE 5.7
N_2 CARS spectra in several temperature and pressure conditions. Shapes of CARS spectra change noticeably depending on the pressure conditions; this feature differs greatly from that of spontaneous Raman spectroscopy.

in CARS spectra, and the shape of CARS spectra changes according to the influence of χ_{nr}. When the concentration of the measured species becomes low, this effect is significantly enhanced and often limits its sensitivity. There are also several effects such as pressure broadening and pressure narrowing, which have to be included into analyses of CARS spectra. As described above, the CARS signal intensity depends on the square of the number density of the Raman-active molecule n^2. Strictly speaking this is valid only in nonoverlapped Doppler broadened transitions. Under pressure-broadening and pressure-narrowing effects, the dependence of the number density varies, ranging between n and n^2 in accordance with the measurement conditions. Figure 5.7 shows N_2 CARS spectra in several temperature and pressure conditions.[5.1],[5.3],[5.7] Shapes of CARS spectra change noticeably depending on the pressure conditions, and this feature greatly differs from that of spontaneous Raman spectroscopy.

There are several factors in obtaining accurate signals using the Raman technique. Table 5.2 summarizes the characteristics of CARS. Several parameters such as pressure broadening and pressure narrowing affect CARS spectra, and it is important to understand these effects correctly.

1. *Background noise.* Since the CARS signal is strong and has a beamlike form, it is easier to block out noises like fluorescence from the CARS signal. On the other hand, CARS generates a background noise in its

TABLE 5.2

Summary of CARS Characteristics

	Characteristics	Countermeasure
Theoretical treatments	• Energy transition • Selection rule • Pressure broadening and pressure narrowing	
Temperature	• Boltzmann distribution (lower energy states)	• Theoretical evaluation
Pressure	• Pressure broadening and pressure narrowing	• Theoretical evaluation
Windows	• Sensitive	• Purge gas • Heating
Calibration	• Medium	• Theoretical evaluation
Noise	• Nonresonant susceptibility	• Short pulse (ps or fs) laser
Measurement item	• Temperature • Concentration	
Measurement dimension	• Point	
Detection limit	• %	• Short pulse (ps or fs) laser
stability	• Laser • Beam steering	• Single-beam CARS

Note: Several parameters such as pressure broadening and pressure narrowing affect CARS spectra, and it is important to understand these effects.

process, which is due to the influence of the nonresonant susceptibility χ_{nr}. χ_{nr} arises in a nonresonant CARS process, and this causes deformation of a resonant CARS signal and a decrease in signal-to-noise ratio. Therefore, the prediction of χ_{nr} behavior before applying the CARS measurement becomes important to reduce measurement risks.

2. *Dirt on measurement windows.* A lack of contamination of measurement windows is crucial for CARS. Since the CARS process depends on phase matching of incident lights, degradation of the incident laser beam quality has a large influence on CARS signals. It is also important to keep windows clear for the measurement of sufficient CARS signals.

3. *Beam steering.* Beam steering is much more important in CARS than with other laser diagnostics. Because CARS is a nonlinear Raman process and uses phase-matching conditions, beam steering for each incident laser must be maintained. The movement of each laser pass means a sharp decrease of the CARS signal intensity. It is also worth noting that the optical pass has to be adjusted under experimental conditions, especially in high pressure.

4. *Pressure effects.* There are several pressure effects in CARS, including pressure-broadening and pressure-narrowing phenomena.

According to these effects, the dependency of CARS signal intensity on the number density n may also change in a complicated way, as discussed above. Therefore, the prediction of these effects using a CARS simulation is necessary for many applications.

5. *Temperature effect and changes of population fraction in energy levels.* The effect of temperature has almost the same tendency as that of spontaneous Raman spectra, and CARS spectra change in shape and intensity by pressure-broadening and pressure-narrowing phenomena, as shown in Figure 5.7. In quantitative concentration and temperature measurements, it is also necessary to predict these effects using the simulation of CARS spectra in the considered temperature and pressure ranges. As mentioned above, pressure plays a key role in CARS spectra, and temperature and pressure must be considered simultaneously.

5.2 Geometric Arrangement and Measurement Species

5.2.1 Spontaneous Raman Spectroscopy

A typical geometric arrangement of spontaneous Raman spectroscopy is shown in Figure 5.8. Both pulsed and continuous lasers are used for spontaneous Raman spectroscopy. The scattering signals from the measurement area are collected and measured by a spectrometer. A charge-coupled device

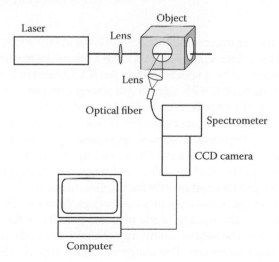

FIGURE 5.8
Typical geometric arrangement of spontaneous Raman spectroscopy. Main components of a Raman spectroscopy system are a laser, a spectrometer, and a CCD camera. A Raman scattering signal is very weak, and sensitive detectors are necessary to get sufficient signals.

(CCD) camera, which often has an amplifying function, is mostly used to detect the Raman signals. Since spontaneous Raman signals are weak, filters for both laser light and Raman scattering are necessary to reduce noise in the Raman-scattering wavelength region.

Applying spontaneous Raman spectroscopy to practical applications, several factors have to be taken into account. Figure 5.9 shows application procedures that have to be considered in spontaneous Raman spectroscopy applications. The most important factor for spontaneous Raman spectroscopy is noise reduction. If there is no emission from the measurement area, a continuous laser and a low-light CCD camera is a preferable system. However, in conditions of noticeable emissions, including flames, a pulsed laser and an image-intensified CCD (ICCD) camera have to be selected. Fluorescence interference can sometimes be reduced using short-pulse lasers. Spontaneous Raman spectroscopy can be used for both the clarification of basic phenomena in industrial processes and the monitoring and advanced control of industrial systems.

Spontaneous Raman spectroscopy can be used for both species concentration and temperature measurements. Typical gas species detected by spontaneous Raman spectroscopy are shown with their Raman shift in Table 5.3.[5.3],[5.8] Solid, liquids, and larger gas species can also be detected by spontaneous Raman spectroscopy. The sensitivity depends on their Raman scattering cross sections, and they are also dependent on the incident light wavelength. The sensitivity of spontaneous Raman spectroscopy is not high, but major species can be detected by spontaneous Raman spectroscopy.

5.2.2 CARS

A typical geometric arrangement of CARS is shown in Figure 5.10. Pulsed lasers are mostly used for CARS. CARS signals form a beamlike feature and they are collected and dispersed by a spectrometer. An ICCD camera is mostly used to detect the CARS signals. CARS signals are strong compared to spontaneous Raman signals, and they are not susceptible to emissions or fluorescence.

When applying CARS to practical applications, several factors have to be taken into account. Figure 5.11 shows application procedures that have to be considered in CARS applications. The most important factors for CARS are to understand its spectra and to stabilize the optical setup (phase-matching conditions). CARS can be used mostly for the clarification of basic phenomena in industrial processes because of its system complication. The CARS signal intensity depends on the square of the number density of the Raman-active molecule. Therefore, the signal intensity strongly depends on the number density of the target molecule. The shape of CARS signals also depends on the number density, as discussed in Section 5.1.2. Therefore, it is sometimes tricky to understand shape-changing CARS signals. These phenomena usually do not happen in laser-induced fluorescence (LIF), spontaneous Raman

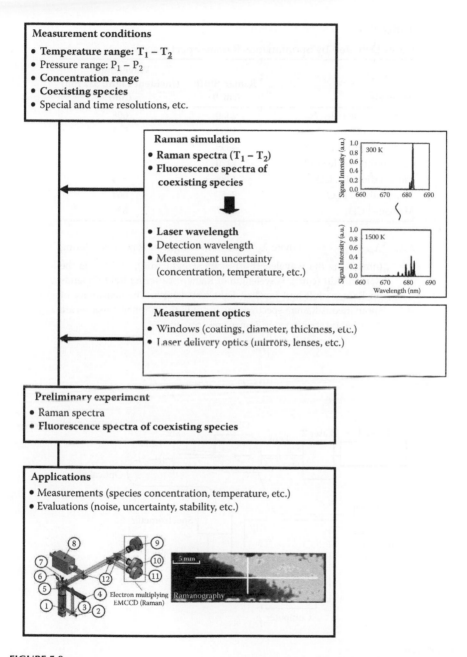

Measurement conditions

- **Temperature range: $T_1 - T_2$**
- Pressure range: $P_1 - P_2$
- **Concentration range**
- **Coexisting species**
- Special and time resolutions, etc.

Raman simulation

- **Raman spectra $(T_1 - T_2)$**
- **Fluorescence spectra of coexisting species**

- **Laser wavelength**
- Detection wavelength
- Measurement uncertainty (concentration, temperature, etc.)

Measurement optics

- Windows (coatings, diameter, thickness, etc.)
- Laser delivery optics (mirrors, lenses, etc.)

Preliminary experiment

- Raman spectra
- **Fluorescence spectra of coexisting species**

Applications

- Measurements (species concentration, temperature, etc.)
- Evaluations (noise, uncertainty, stability, etc.)

FIGURE 5.9

Application procedures in spontaneous Raman spectroscopy. The most important factor for spontaneous Raman spectroscopy is noise reduction. Fluorescence interference can sometimes be reduced using short-pulse lasers.

TABLE 5.3

Gases Detected by Spontaneous Raman Spectroscopy

Molecules	Raman Shift (cm^{-1})	Wavelength (Incident Light: 532 nm) (nm)
Nitrogen molecule : N_2	2360	608
Oxygen molecule: O_2	1580	581
Hydrogen molecule : H_2	4395	694
Carbon monoxide : CO	2170	601
Carbon dioxide : CO_2	1388	574
Water vapor : H_2O	3657	661
Methane : CH_4	2917	630

Note: $\lambda_M = \left[\dfrac{10^7}{10^7/\lambda_0 - \overline{v}_M} \right]$, where λ_M is the wavelength of Raman scattering (nm), λ_0 is the wavelength of incident light (nm), and $\overline{v}_M$ is the Raman shift (cm^{-1}). Wavelength of Raman scattering light depends on Raman shift and incident light wavelength. The sensitivity of spontaneous Raman spectroscopy is not high, and major species are detected by spontaneous Raman spectroscopy.

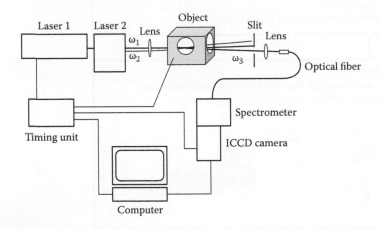

FIGURE 5.10

Typical geometric arrangement of CARS. In CARS two laser systems are usually employed for pump and probe beams. They are crossed with each other and a beamlike CARS signal is generated in the phase-matching condition.

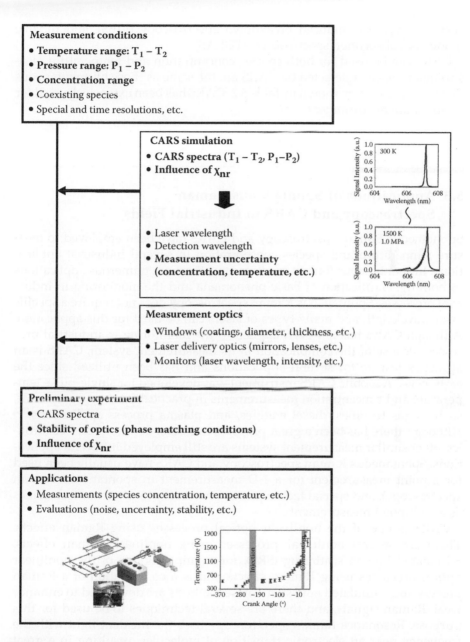

FIGURE 5.11

Application procedures in CARS. The most important factors for CARS are to understand its spectra and to stabilize the optical setup. CARS can be used mostly for the clarification of basic phenomena in industrial processes because of its system complication. The CARS signal intensity depends on the square of the number density of the Raman-active molecules. Therefore, the signal intensity strongly depends on the number density of the target molecule.

spectroscopy, laser-induced breakdown spectroscopy (LIBS), and tunable diode laser absorption spectroscopy (TDLAS).

CARS can be used for both species concentration and temperature measurements. Species detected by CARS are the same as those by spontaneous Raman spectroscopy shown in Table 5.2. CARS has been extensively used for temperature measurement.

5.3 Applications of Spontaneous Raman Spectroscopy and CARS to Industrial Fields

Spontaneous Raman spectroscopy and CARS have been employed to measure temperature and species concentration in several industrial applications. In spontaneous Raman spectroscopy, there are numerous applications in both the clarification of basic phenomena and the monitoring in industrial processes. Spontaneous Raman spectroscopy does not require a specific laser wavelength and many types of lasers can be used for this application. Although CARS has not been applied to the monitoring in industrial processes because of the complication in its measurement system, CARS is an attractive feature for several applications and has been utilized since the early 1980s. A mobile CARS instrument was developed for single-pulse temperature and concentration measurements in practical combustion systems, such as gas turbines, diesel engines, and plasma process applications.[5.9] Although there has been a great progress in lasers, detectors, and electronics, other similar measurement systems are still employed in several applications. Spontaneous Raman spectroscopy and CARS have usually been used for a point measurement (or a 1-D measurement in spontaneous Raman spectroscopy), and special techniques are necessary to use these techniques as a multipoint measurement.

CARS is one of the nonlinear optical processes using Raman effects. There are several nonlinear processes using nonlinear Raman effects. Stimulated Raman scattering (SRS), for example, is one of the nonlinear optical processes using Raman effects and is a combination of a Raman process and stimulated emission. These methods are developed to enhance weak Raman signals, and there are several techniques to be used for this purpose. Resonance Raman spectroscopy uses the incident light with the frequency near an electronic transition of molecules, resulting in a great enhancement of Raman scattering intensity. Surface-enhanced Raman spectroscopy (SERS) is a method that enhances Raman scattering intensities of molecules on metal surfaces like Ag. The enhancement can be as much as 10^{10} or more, and this technique has the capability to detect a single molecule.[5.10] Though these methods work well in some conditions, complication of the signal-generation processes sometimes makes their signals susceptible to many parameters. If spontaneous Raman spectroscopy can be applicable to industrial needs, it is the recommended method for them.

It is worth noting that there have been great and rapid advances of these technologies in the life sciences field. For example, they have been used for the molecular identification inside live cells, and CARS has great potential to subdue noises including fluorescence. Picosecond and femtosecond CARS have been actively used in this area. This is briefly described in Chapter 8.

5.3.1 Engine Applications

Both spontaneous Raman spectroscopy and CARS have been used for engine research. CARS was an excellent tool for temperature measurement[5.11],[5.12] and was actively employed for in-cylinder temperature measurements in 1980s and 1990s. However, there are now several methods to be employed for temperature and concentration measurements in engine combustion, such as LIF and TDLAS. These methods are now more frequently used for engine combustion analyses.

There is a special advantage for spontaneous Raman spectroscopy and CARS compared to LIF and TDLAS. Both spontaneous Raman spectroscopy and CARS can measure the hydrogen molecule (H_2). Though H_2 is an important fuel in combustion, neither LIF nor TDLAS can detect H_2, while spontaneous Raman spectroscopy and CARS have a good sensitivity for H_2. Because of this feature, both methods have often been applied to hydrogen engine combustion.[5.13],[5.14] Fluorescence interference from large hydrocarbons such as polycyclic aromatic hydrocarbons (PAHs) does not arise in H_2 combustion, resulting in a preferable condition for spontaneous Raman spectroscopy. Figure 5.12 shows simultaneous temperature and exhaust-gas recirculation measurements in a homogeneous charge-compression ignition engine by pure rotational CARS.[5.14] Using a pure rotational CARS spectra, O_2 and N_2 CARS spectra can be detected simultaneously. Temperature and the amount of recirculated exhaust gas (relative O_2 and N_2 concentrations) can be determined from these CARS spectra.

Usually spontaneous Raman spectroscopy has been used for a point or 1-D measurement. However, in case of low-noise conditions, a 2-D measurement becomes possible using an appropriate filter and a CCD detector. Figure 5.13 shows a measurement result of 2-D H_2 concentration by spontaneous Raman spectroscopy.[5.15] The simultaneous measurements of LIF and spontaneous Raman scattering were applied to a hydrogen internal combustion (IC) engine. Triethylamine (TEA) was used as a tracer molecule for the LIF experiments. The Raman and LIF results were found to be in good agreement with each other.

5.3.2 Gas Turbine Applications

Gas turbines have a reasonable scale for laser diagnostics and have been studied using both spontaneous Raman spectroscopy and CARS. Figure 5.14

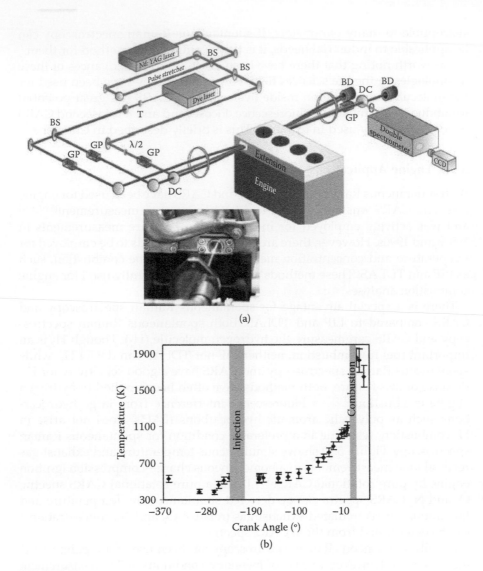

(a)

(b)

FIGURE 5.12

Temperature and species concentration measurements using CARS. (a) Experimental setup for dual-broadband pure rotational CARS; *BS:* beam splitter, *T:* telescope, *GP:* Glan–Taylor prism, *λ/2:* half-wave plate, *DC:* dichroic mirror, and *BD:* beam dump. (b) CARS measured temperatures before autoignition and after combustion. The points represent mean values of single-pulse measurements and error bars represent the standard deviation. Gray bars denote crank-angle degrees where no measurements were possible. (Reprinted from M.C. Weikl, F. Beyrau, and A. Leipertz, "Simultaneous temperature and exhaust-gas recirculation-measurements in a homogeneous charge-compression ignition engine by use of pure rotational coherent anti-Stokes Raman spectroscopy," *Applied Optics*, 45(15), 3646–3651, 2006. With permission of Optical Society of America.)

shows one of the applications of spontaneous Raman spectroscopy to gas turbines.[5,16] The burner can produce up to a thermal power of 370 kW at 0.3 MPa using natural gas as a fuel. The air preheat temperature was kept to 673 K. Single-shot 1-D Raman measurements were applied to this burner for quantitative measurements. Time-averaged and single-shot OH were also detected for a qualitative analysis of the position and shape of the flame brush, the flame front, and the stabilization mechanism (see Chapter 3). Figure 5.13(a) shows the experimental setup in this experiment. The Nd:YAG laser at 532 nm was used as a light source and an ICCD camera as a detector. N_2, O_2, H_2O, CO_2, and H_2 Raman signals are detected to measure major

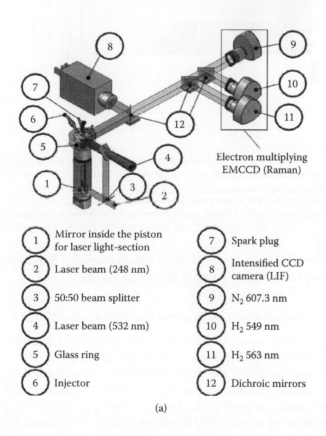

1 Mirror inside the piston for laser light-section	7 Spark plug
2 Laser beam (248 nm)	8 Intensified CCD camera (LIF)
3 50:50 beam splitter	9 N_2 607.3 nm
4 Laser beam (532 nm)	10 H_2 549 nm
5 Glass ring	11 H_2 563 nm
6 Injector	12 Dichroic mirrors

(a)

FIGURE 5.13a
Measurement result of 2-D H_2 concentration by spontaneous Raman spectroscopy. Optical accessible engine and optical setup for LIF/Raman measurements. Optical excitation pathways are shown for LIF and spontaneous Raman spectroscopy. (Reprinted from A. Braeuer, and A. Leipertz, "Two-dimensional Raman mole-fraction and temperature measurements for hydrogen–nitrogen mixture analysis," *Applied Optics*, 48(4), B57–B64, 2009. With permission of Optical Society of America.)

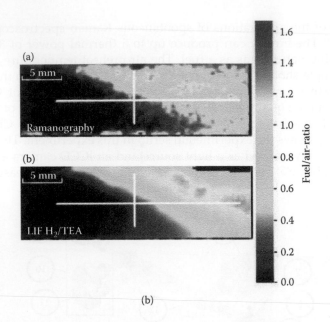

(b)

FIGURE 5.13b
Measurement result of 2-D H_2 concentration by spontaneous Raman spectroscopy. Fuel-air ratio distribution inside the combustion chamber of a hydrogen IC engine. (a) Raman imaging and (b) TEA-LIF imaging averaged over 16 hydrogen jet injections. Measurements were carried out $-42°$ CA before firing TDC. (Reprinted from A. Braeuer, and A. Leipertz, "Two-dimensional Raman mole-fraction and temperature measurements for hydrogen–nitrogen mixture analysis," *Applied Optics*, 48(4), B57–B64, 2009. With permission of Optical Society of America.)

species concentration and temperature simultaneously. Radial profile of the mean mixture fraction in a flame is shown in Figure 5.13(b).

5.3.3 Online Monitoring and Process Control Applications

Spontaneous Raman spectroscopy has a lot of excellent features for on-line monitoring. Though spontaneous Raman spectroscopy can be applicable to solid, liquid, and gas phase materials, applications to solid and liquid materials are dominant in on-line monitoring. Number densities of solid and liquid materials are much larger than that of gas, and applications of spontaneous Raman spectroscopy to solid and liquid materials can mitigate the drawback of weak signal intensities. Applications of spontaneous Raman spectroscopy include polymerization processes,[5.17] pharmaceutical processes,[5.18]–[5.20] freeze-drying processes,[5.21] thin film productions,[5.22] and so on. It is worth noting that crystallization can be monitored by spontaneous Raman spectroscopy because bonding states of atoms and molecules make an impact on Raman spectra. Most of these applications have used simple and low-cost Raman spectrometers, which is one of the great merits of spontaneous Raman spectroscopy.

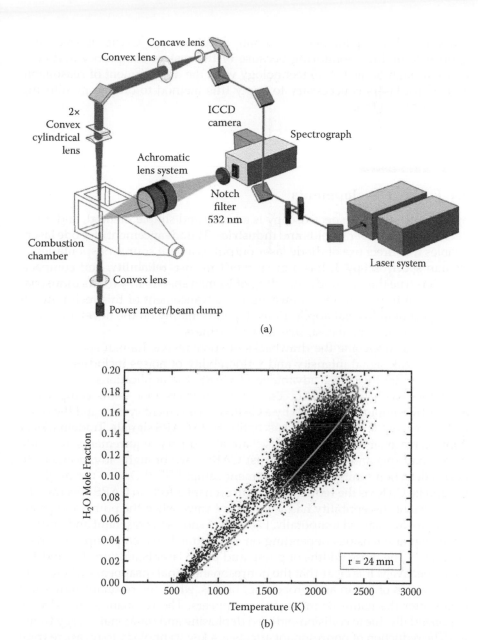

(a)

(b)

FIGURE 5.14

Spontaneous Raman spectroscopy application to the gas turbine combustor. (a) Schematic drawing of the setup for 1-D Raman; (b) Scatter plot of H_2O versus temperature. Scatter plot of H_2O versus temperature in flame A at a radius of r = 24 mm, 22 mm downstream from the burner mouth. The light gray line indicates the adiabatic equilibrium state. (Reprinted from H. Ax, U. Stopper, W. Meier, M. Aigner, and F. Güthe, *Journal of Engineering for Gas Turbines and Power*, 132(5), 051503/1, 2010. With permission from American Society of Mechanical Engineers.)

Compared to spontaneous Raman spectroscopy, CARS hasn't been applied to on-line monitoring because of the system complexity and cost. Breakthrough of the CARS technology with the development of reasonable and robust lasers is necessary to apply this method to on-line monitoring and process control.

5.4 Future Developments

Spontaneous Raman spectroscopy is an easy and simple method, and it has been applied in several fields and industries. The advancement of diode lasers enables the direct use of diode laser output as a light source of spontaneous Raman spectroscopy. It has a great merit in cost, reliability, and compactness. It is true that the understanding of Raman spectra in various industrial fields is an important task; however, the advancement of this technique in the aspect of industrial applications depends mainly on the development of lasers and detectors in cost, size, and robustness.

CARS has overcome the drawbacks of spontaneous Raman spectroscopy, such as weak signal intensity and vulnerability of noises including fluorescence. Despite this great advantage of CARS, the applications of CARS to industries has had limited success for two reasons. One is the complicated and costly equipment required by a CARS measurement system and the other is the effect of the nonresonant susceptibility to CARS signals. Trademarks of "simple, reasonable, and easy-to-use" are all industry requirements. Recently picosecond (ps) and femtosecond (fs) CARS have brought advancement to CARS detection limits and measurement setups.[5.7],[5.28] One of the merits of ps- and fs- CARS is the reduction of nonresonant effects in CARS spectra. The nonresonant susceptibility has a significant value when the pump and probe beams are overlapped temporally. Figure 5.15 shows the resonant and nonresonant signal intensities depending on delay time between pump and probe beams.[5.24] The pulse widths of pump and probe laser beams are 135 and 106 ps, respectively. It is clear that the nonresonant signal disappears when there is no overlap of pump and probe laser beams, while the resonant signal still exists after the nonresonant signal disappears. The resonant signal decays exponentially due to collision-induced dephasing and rotational energy transfer. The reduction of nonresonant effects is a key technology to measure trace species by CARS, and it will pave the way for the further CARS applications. It is also worth noting that the CARS signal can be detected using a single-beam CARS configuration. This method reduces the complexity and vulnerability of optical setups in CARS. Figure 5.16 shows one of the single-beam CARS applications.[5.25] Trace amounts of solids are detected by CARS using shaped fs pulses. Compact ps- and fs-lasers are now available and further advances of these lasers will bring CARS to various industrial applications.

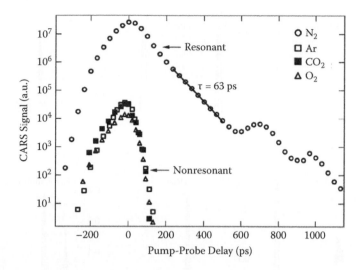

FIGURE 5.15

Resonant and nonresonant signal intensities depending on delay time between pump and probe beams. Time-resolved resonant and nonresonant signals as functions of pump–probe time delay displaying the behavior of the resonant and nonresonant CARS signal. (Reprinted with permission from S. Roy, T.R. Meyer, and J.R. Gord, "Time-resolved dynamics of resonant and nonresonant broadband picosecond coherent anti-Stokes Raman scattering signals," *Applied Physics Letters*, 87(26), 264103/1–264103/3, Copyright 2005. American Institute of Physics.)

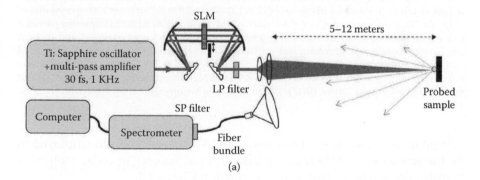

FIGURE 5.16a

Single-beam CARS application for trace amounts of solids detection. Temperature measurement results. The laser source is an amplified Ti:sapphire laser with a dispersion compensating prism compressor. The pulses 0.5 mJ, 30 fs at 1 KHz repetition rate are phase shaped in a pulse shaper using an electronically controlled liquid-crystal spatial light modulator. The short-wavelength end of the spectrum is suppressed using an LP interference filter and by using a variable knife-edge slit at the shaper's Fourier plane. The beam is focused on the distant sample through a telescope, and the scattered radiation is collected with a 7.5 in. diameter lens. The collected light is short-pass filtered and is fiber coupled to an imaging spectrometer. (Reprinted with permission from O. Katz, A. Natan, Y. Silberberg, and S. Rosenwaks, "Standoff detection of trace amounts of solids by nonlinear Raman spectroscopy using shaped femtosecond pulses," *Applied Physics Letters*, 92(17), 171116/1–171116/3, Copyright 2008. American Institute of Physics.)

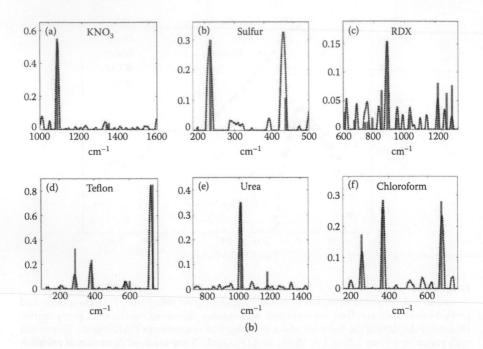

(b)

FIGURE 5.16b

Single-beam CARS application for trace amounts of solids detection, temperature measurement results. Resolved femtosecond CARS vibrational spectra of several scattering samples dashed, obtained at stand-off distances of 12 m [(a)–(c)] and 5 m [(d)–(f)]. (a): 1000 g crystallized KNO₃, (b) 500 g sulfur powder, (c) Cyclotrimethylene-trinitramine RDX/T4 explosive particles with a total mass of 4 mg, (d) bulk PTFE, (e) 4 mg of crystallized urea particles, and (f) 1 cm long cuvette containing chloroform and scattering ZnTe particles 200 nm diameter. (Reprinted with permission from O. Katz, A. Natan, Y. Silberberg, and S. Rosenwaks, "Standoff detection of trace amounts of solids by nonlinear Raman spectroscopy using shaped femtosecond pulses," *Applied Physics Letters*, 92(17), 171116/1–171116/3, Copyright 2008. American Institute of Physics.)

Spontaneous Raman spectroscopy and CARS have also been employed in the life sciences[5.29],[5.30] because of their unique features in cellar molecular identifications, which are briefly described in Chapter 8.

References

[5.1]   A.C. Eckbreth, *Laser Diagnostics for Combustion Temperature and Species*, Cambridge, Mass.: ABACUS Press, 1988.

[5.2]   K. Kohse-Hoinghaus and J.B. Jeffries, *Applied Combustion Diagnostics*, New York: Taylor and Francis, 2002.

[5.3]   G. Herzberg, *Molecular Spectra and Molecular Structure: Spectra of Diatomic Molecules*, New York, Van Nostrand Reinhold, 1950.

[5.4]   B. Zilles and R. Carter, "Computer program to simulate Raman scattering," NASA Contract. Rep. (NASA-CR-145151), 1977.

[5.5]   Y.Y. Kwan, "The interacting states of an asymmetric top molecule XY2 of the group C2v. Application to five interacting states (101), (021), (120), (200), and (002) of water (H216O)." *Journal of Molecular Spectroscopy,* 71, 260–280, 1978.

[5.6]   S. Roy, J.R. Gord, A.K. Patnaik, "Recent advances in coherent anti-Stokes Raman scattering spectroscopy: Fundamental developments and applications in reacting flows," *Progress in Energy and Combustion Science,* 36, 280–306, 2010.

[5.7]   L.A. Rahn and R.E. Palmer, "Studies of nitrogen self-broadening at high temperature with inverse Raman spectroscopy," *Journal of the Optical Society of America B,* 3(9), 1164–1169, 1986.

[5.8]   G. Herzberg, *Molecular Spectra and Molecular Structure: Spectra of Polyatomic Molecules,* New York, Van Nostrand Reinhold, 1966.

[5.9]   T.J. Anderson, G.M. Dobbs, and A.C. Eckbreth, "Mobile CARS instrument for combustion and plasma diagnostics," *Applied Optics,* 25(22), 4076–4085, 1986.

[5.10] B. Pettinger, "Single-molecule surface- and tip-enhanced Raman spectroscopy," *Molecular Physics,* 108(16), 2039–2059, 2010.

[5.11] A.C. Eckbreth, G.M. Dobbs, J.H. Stufflebeam, and P.A. Tellex , "CARS temperature and species measurements in augmented jet engine exhausts," *Applied Optics,* 23(9), 280–306, 1984.

[5.12] S. Roy, T.R. Meyer, R.P. Lucht, V.M. Belovich, E. Corporan, and J.R. Gord, "Temperature and CO_2 concentration measurements in the exhaust stream of a liquid-fueled combustor using dual-pump coherent anti-Stokes Raman scattering (CARS) spectroscopy," *Combustion and Flame,* 138(3), 273–306, 2004.

[5.13] S.R. Engel, P. Koch, A. Braeuer, and A. Leipertz, "Simultaneous laser-induced fluorescence and Raman imaging inside a hydrogen engine," *Applied Optics,* 48(35), 6643–6650, 2009.

[5.14] M.C. Weikl, F. Beyrau, and A. Leipertz, "Simultaneous temperature and exhaust-gas recirculation-measurements in a homogeneous charge-compression ignition engine by use of pure rotational coherent anti-Stokes Raman spectroscopy," *Applied Optics,* 45(15), 3646–3651, 2006.

[5.15] A. Braeuer and A. Leipertz, "Two-dimensional Raman mole-fraction and temperature measurements for hydrogen–nitrogen mixture analysis," *Applied Optics,* 48(4), B57–B64, 2009.

[5.16] H. Ax, U. Stopper, W. Meier, M. Aigner, and F. Güthe, "Experimental analysis of the combustion behavior of a gas turbine burner by laser measurement techniques," *Journal of Engineering for Gas Turbines and Power,* 132(5), 051503/1–051503/9, 2010.

[5.17] T.R. McCaffery and Y.G. Durant, "Application of low-resolution Raman spectroscopy to online monitoring of miniemulsion polymerization," *Journal of Applied Polymer Science,* 86(7), 1507–1515, 2002.

[5.18] N.A. Macleod and P. Matousek, "Emerging non-invasive Raman methods in process control and forensic applications," *Pharmaceutical Research,* 25(10), 2205–2215, 2008.

[5.19] G. Fevotte, "In situ Raman spectroscopy for in-line control of pharmaceutical crystallization and solids elaboration processes: A review," *Chemical Engineering Research and Design,* 85(A7), 906–920, 2007.

[5.20] T.R.M. De Beer, C. Bodson, B. Dejaegher, B. Walczak, P. Vercruysse, A. Burggraeve, A. Lemos, L. Delattreb, Y. Vander Heyden, J.P. Remon, C. Vervaet, and W.R.G. Baeyens, "Raman spectroscopy as a process analytical technology (PAT) tool for the *in-line* monitoring and understanding of a powder blending process," *Journal of Pharmaceutical and Biomedical Analysis*, 48(3), 772–779, 2008.

[5.21] T.R.M. De Beer, M. Allesø, F. Goethals, A. Coppens, Y. Vander Heyden, H. Lopez de Diego, J. Rantanen, F. Verpoort, C. Vervaet, J.P. Remon, and W.R.G. Baeyens, "Implementation of a process analytical technology system in a freeze-drying process using Raman spectroscopy for in-line process Monitoring," *Analytical Chemistry*, 79(21), 7992–8003, 2007.

[5.22] J. Palm, S. Jost, R. Hock, and V. Probst, "Raman spectroscopy for quality control and process optimization of chalcopyrite thin films and devices," *Thin Solid Films*, 515(15), 5913–5916, 2007.

[5.23] M.R. Leahy-Hoppa, J. Miragliotta, R. Osiander, J. Burnett, Y. Dikmelik, C. McEnnis, and J.B. Spicer, "Ultrafast laser-based spectroscopy and sensing: Applications in LIBS, CARS, and THz spectroscopy," *Sensors*, 10, 4342–4372, 2010.

[5.24] S. Roy, T.R. Meyer, and J.R. Gord, "Time-resolved dynamics of resonant and nonresonant broadband picosecond coherent anti-Stokes Raman scattering signals," *Applied Physics Letters*, 87(26), 264103/1–264103/3, 2005.

[5.25] O. Katz, A. Natan, Y. Silberberg, and S. Rosenwaks, "Standoff detection of trace amounts of solids by nonlinear Raman spectroscopy using shaped femtosecond pulses," *Applied Physics Letters*, 92(17), 171116/1–171116/3, 2008.

[5.26] T. R. Meyer, S. Roy, and J. R. Gord, "Improving Signal-to-Interference Ratio in Rich Hydrocarbon-Air Flames Using Picosecond Coherent Anti-Stokes Raman Scattering," *Appl. Spectrosc.* 61(11), 1135–1140, 2007.

[5.27] H. U. Stauffer, W. D. Kulatilaka, P. S. Hsu, J. R. Gord, and S. Roy, "Thermometry of gas-phase reacting flows using time-delayed picosecond coherent anti-Stokes Raman scattering spectra of H_2," *Applied Optics* 50, A38–A48, 2011.

[5.28] W. D. Kulatilaka, P. S. Hsu, H. U. Stauffer, J. R. Gord, and S. Roy, "Direct measurement of rotationally resolved H2 Q-branch Raman coherence lifetimes using time-resolved picosecond CARS," *Applied Physics Letters* 97, 081112, 2010.

[5.29] R.J. Swain and M.M. Stevens, "Raman microspectroscopy for non-invasive biochemical analysis of single cells," *Biochemical Society Transactions*, 35(3), 544–549, 2007.

[5.30] S.H. Kim, E.-S. Lee, J.Y. Lee, E.S. Lee, B.-S. Lee, J.E. Park, and D.W. Moon, "Multiplex coherent anti-Stokes Raman spectroscopy images intact atheromatous lesions and concomitantly identifies distinct chemical profiles of atherosclerotic lipids," *Circulation Research*, 106(8), 1332–1341, 2010.

6

Tunable Diode Laser Absorption Spectroscopy

6.1 Principle

Tunable diode laser absorption spectroscopy (TDLAS) uses the absorption phenomena to measure species concentration and temperature. When light permeates an absorption medium as shown in Figure 6.1, the molecular concentration is in proportion to the strength of the transmitted light according to the Lambert Beer's law. Atomic or molecular concentration is related to the amount of light absorbed, as in the following formula:[6.1],[6.2]

$$I_\lambda/I_0 = \exp(-\kappa n \ell)$$
$$= \exp(-\alpha) \tag{6.1}$$

Here, I_0 is the input laser light intensity, I_λ the transmitted light intensity at wavelength λ, κ the absorption coefficient, n the species number density, and ℓ the path length. The absorption coefficient κ is related to the line strength of the absorption transition. Thus, by measuring the attenuation of light that permeates an absorption medium containing atoms or molecules, their concentration can be ascertained. As seen in Equation (6.1), the absorption I_λ/I_0 is not directly proportional to the species number density n. The absorbance $\ln(I_0/I_\lambda)$ is often defined as

$$\alpha = \ln(I_0/I_\lambda) \tag{6.2}$$

The absorbance is directly proportional to the species number density n and also the path length ℓ. Cautious use of the absorbance $\ln(I_0/I_\lambda)$ is necessary because $\log_{10}(I_0/I_\lambda)$ is usually used in analytical chemistry. Using the natural logarithm instead of the common logarithm to define the absorbance is rather limited in the TDLAS applications.

Equation (6.1) is a simplified form of the absorption theory, and there are several terms that must be included in Equation (6.1). One of them is the line

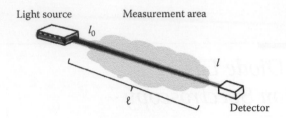

FIGURE 6.1
Light transmission through an absorption medium. When light transmits through an absorption medium, the strength of the transmitted light is related to absorber concentration according to Lambert Beer's law. Atoms and molecules have their own spectral pattern. Because of
this feature, TDLAS has excellent selectivity and sensitivity.

broadening effect of absorption lines. There are three types of line broadenings, which are described in Appendix C. Usually the natural broadening is
small and does not have a considerable contribution to actual spectra observed
in practical applications. The Doppler and collision broadenings are dominant in practical applications, and they have the line shape functions[6.3]

$$G_D(\nu) = \frac{c}{\nu_0} \sqrt{\frac{m}{2\pi kT}} \exp\left[-4\ln 2 \cdot \frac{(\nu - \nu_0)^2}{\nu_D^2}\right] \tag{6.3}$$

$$G_C(\nu) = \frac{\nu_C}{2\pi} \frac{1}{(\nu - \nu_0)^2 + \left(\frac{\nu_C}{2}\right)^2} \tag{6.4}$$

where $G_D(\nu)$ and $G_C(\nu)$ are the Doppler and collision broadenings respectively,
ν is the frequency of light, ν_0 the transition center frequency, c the speed of
light, m the atomic or molecular mass, k the Boltzmann constant, and $\Delta\nu$ the
transition full width at half maximum (FWHM). $\Delta\nu_D$ and $\Delta\nu_C$ are FWHMs of
the Doppler and collision broadenings. The combination of the Doppler and
collision broadenings is described by the Voigt function as

$$G_V(a, x) = \frac{a}{\pi} \int_{-\infty}^{\infty} \frac{e^{-y^2}}{a^2 + (x - y)^2} dy \tag{6.5}$$

where a and x are defined as

$$a = \sqrt{\ln 2} \frac{\nu_C}{\nu_D} \tag{6.6}$$

$$x = \sqrt{\ln 2} \frac{(\nu - \nu_0)}{\nu_D} \tag{6.7}$$

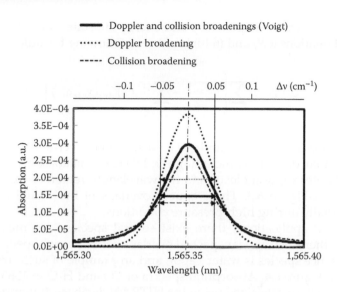

FIGURE 6.2
Line broadenings. The natural broadening is small and does not have considerable contribution to actual spectra observed in practical applications. The Doppler and collision broadenings are dominant in practical applications. The combination of the Doppler and collision broadenings is described by the Voigt function. The Voigt function is easily calculated and often used for modeling absorption and emission spectra.

These broadening shapes are shown in Figure 6.2. The line shape functions are normalized to form the following relationship:

$$\int G(v)dv = 1 \tag{6.8}$$

The line shape of absorption spectra is important to evaluate the quantitative species concentrations. With respect to the absorption lines in the wavelength region observed, Equation (6.1) can be given by

$$\alpha(v) = \sum_i \left(n_i \ell \sum_j S_{i,j}(T) G_{V,j}(x_{i,j}, a_{i,j}) \right) \tag{6.9}$$

where T is the temperature, and $S_{i,j}$ the line strength of the transition from i to j states. n_i is the number density of the atom or molecule at the i state and is described by the Boltzmann equation

$$n_i = n \frac{g_i e^{-E_i/kT}}{\sum_j g_j e^{-E_j/kT}}$$

$$= n \frac{g_i e^{-E_i/kT}}{Z} \tag{6.10}$$

where g_i and E_i are the degeneracy and energy of i state, and Z the partition function. Equations (6.9) and (6.10) lead to the following formula:

$$\alpha(v) = n\ell \sum_i \left(\frac{g_i e^{-E_i/kT}}{Z} \sum_j S_{i,j}(T) G_{V,j}(x_{i,j}, a_{i,j}) \right) \tag{6.11}$$

Using Equations (6.1), (6.10), and (6.11), absorption spectra are synthesized with molecular databases, including the HITRAN database.[6.1] Absorption spectra of water vapor molecules in the wavelength region of 1350–1370 nm are shown in Figure 6.3. There appear overlaps of individual absorption lines, especially during high-pressure conditions.

In practical applications, there exist several species in a measurement region and there also appear spectral overlaps between each species. A typical interference species is water vapor, and an example of such interference is shown in Figure 6.4. Absorption spectra of CO and H_2O in 1565–1567 nm are theoretically synthesized using the HITRAN database. It is apparent that the CO absorption spectrum has significant interference with those of H_2O. Absorption spectra are often calculated theoretically with precision, and it is important to select the lines having no or less interference with other molecules. These are called "isolated lines," and the marked absorption line of CO in Figure 6.4(a) is preferable to measure CO in coexistence with H_2O. Interference of H_2O strongly depends on temperature and pressure. It is often important to do theoretical screening of measurement species in measurement conditions before applying TDLAS to practical fields.

Table 6.1 shows absorption spectra of H_2O, CO, CO_2, NO, and N_2O synthesized using HITRAN. The best way to evaluate absorption characteristics is using databases that include HITRAN. It is worth noting that all the absorption lines are not always included in the databases; literature checks are also necessary to confirm the validity of the simulation results. There are several key factors in obtaining low-noise signals by TDLAS. The method is theoretically simple, and it is important to reduce noise as much as possible to detect trace absorption signals. These noise effects depend largely on lasers and optics, and careful consideration is necessary when selecting these components.

1. *Noise from lasers and background noise.* There are several methods to reduce these types of noises. The typical methods are 2f modulation and balancing techniques, which are described in Section 6.2. Most of the TDLAS systems employ the self-calibration method, which is not susceptible to emission from a measurement field.

2. *Etalon effects.* An etalon effect arises if there are two parallel surfaces in a laser path, and sine-wave noises are added on the absorption signals. The effect is unique to TDLAS, and it is often called "fringe"

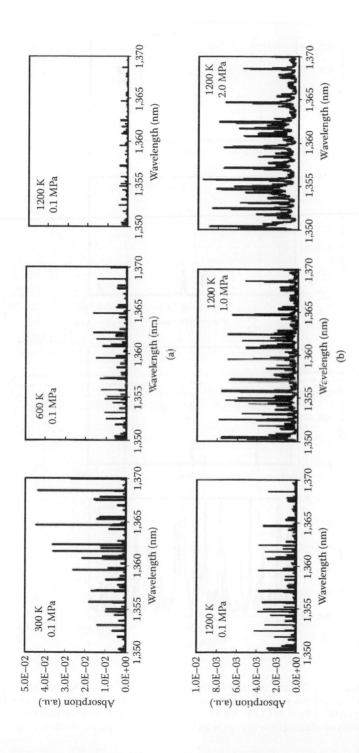

FIGURE 6.3

Absorption spectra of H_2O in the wavelength region of 1350–1370 nm. Absorption spectra of H_2O are calculated using the HITRAN database at the temperature range of 300–1200 K and pressure 0.1–2.0 MPa. There appear overlaps of individual absorption lines, especially in high-pressure conditions.

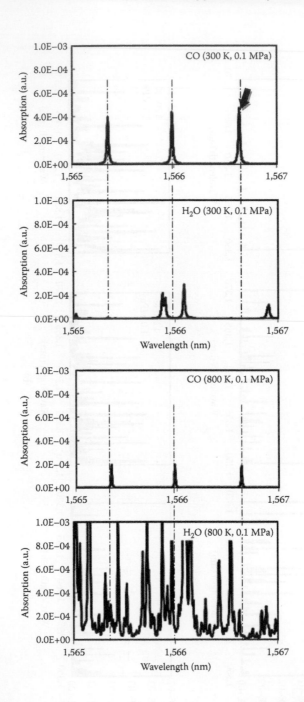

FIGURE 6.4
Absorption spectra of (a) CO and (b) H_2O in 1565–1567 nm. Absorption spectra of CO and H_2O in 1565–1567 nm are theoretically synthesized in different temperature conditions. It is possible to select isolated lines of CO at 300 K. However, it becomes difficult to choose a CO absorption line without causing interference with H_2O absorption spectra at 800 K.

TABLE 6.1

Characteristics of TDLAS

	Characteristics	Countermeasure
Theoretical treatments	• Absorption	
Temperature	• Boltzmann distribution (lower energy states)	• Theoretical evaluation
Pressure	• Collision broadening	• Theoretical evaluation
Windows	• Insensitive	• Purge gas • Heating
Calibration	• Easy	• Self-calibration
Noise	• Interference with other atomic or molecular absorptions • Etalon effect	• Theoretical evaluation • Wedged and coated optics
Measurement item	• Temperature • Concentration (velocity, pressure)	
Measurement dimension	• Line of sight (2-D)	• CT
Detection limit	• ppb–ppm	• Multipass cell
Stability	• Beam steering	

Note: The method is theoretically simple but it is important to reduce noise as much as possible to detect trace absorption signals. These noise effects depend largely on lasers and optics, and careful consideration is necessary to select these components.

in absorption spectroscopy. It is the most common noise in TDLAS and often arises if parallel windows are used as optical components. Wedged windows and antireflection coatings are remedies to reduce these effects. It can also appear from surfaces of optical fibers; A-PC (angled-PC) fibers are recommended to reduce the reflection from fiber surfaces.

3. *Effects of beam steering.* As shown in Equation (6.1), the absorption intensity depends on the path length, and it is often necessary to make a measurement system with more than 1 m path length. Therefore techniques of stable beam steering are the key to achieving a robust measurement system.

4. *Interference from other atomic or molecular absorptions.* As described above, interferences with other atomic or molecular absorptions are the main reason for lower detection limits of TDLAS. Therefore the careful selection of absorption lines—i.e., laser wavelengths—is necessary for practical applications.

5. *Temperature effects and changes of population fraction at energy levels.* The population fraction at each molecular energy level is dependent on temperature by the Boltzmann equation. In quantitative concentration measurements, if the temperature is not measured

simultaneously, it is necessary to select the absorption line so that the population fraction of the lower energy level is constant within a considered temperature range.

6. *Pressure effects.* Because of the collision broadening process, absorption spectra become broader as pressure increases, and broader absorption spectra result in the mixing of absorption lines. These effects complicate the understanding of the absorption signal and also increase noise in absorption signals. Broader absorption spectra have another adverse effect for the signal detection. The signal analyses of TDLAS are often based on the frequency modulation methods. As discussed in Section 6.2, broadening of absorption lines degrades the detection limit of this method.

7. *Dirt on measurement windows.* The surface cleanliness of measurement windows is an important factor in TDLAS; however, this effect is less critical than with other laser diagnostics. Dirt on windows causes several troubles such as attenuation and scattering of the incident light. The self-calibration method of TDLAS makes this method insusceptible to attenuation of laser intensity. This feature makes the method more attractive in practical applications than other laser diagnostics.

6.2 Geometric Arrangement and Measurement Species

A typical geometric arrangement of TDLAS is shown in Figure 6.5(a). As can be expected from its name, tunable diode lasers are used as a light source, and a laser light transmits through the measurement field. The transmitted light is measured by a photodiode. As shown in Figure 6.5(b), wedged windows are commonly used for the laser access to reduce etalon effects. The laser light is usually modulated in wavelength to enhance the detectability of absorption signals. Distributed feedback (DFB) lasers are most frequently used for various applications, and others include distributed Bragg reflector (DBR) lasers, vertical cavity surface emitting lasers (VCSELs), and external-cavity diode lasers (ECDLs). However, photodiodes are a common detector. The structures and features of photo-diodes are shown in Chapter 1. TDLAS has been used for both the clarification of basic phenomena in industrial processes and the monitoring and advanced control of industrial systems. When TDLAS is used for the practical applications, there are several key factors as shown in Section 6.1. Figure 6.6 shows application procedures that have to be considered in TDLAS applications. The most important factors for TDLAS are the selection of laser wavelength and the reduction of noise. The theoretical predictions of TDLAS spectra are important when selecting laser wavelength in practical applications.

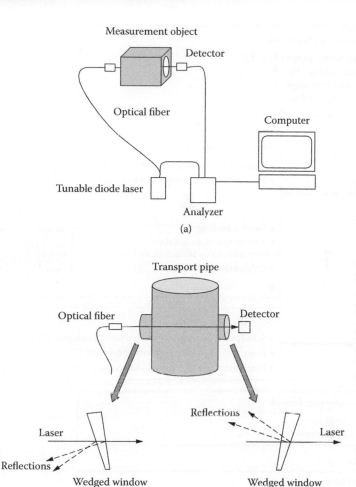

FIGURE 6.5
Typical geometric arrangement of TDLAS. Tunable diode lasers are used as a light source, and a laser light transmits through the measurement field. The transmitted light is measured by a photodiode. Wedged windows are commonly used for the laser access to reduce etalon effects. The laser light is usually modulated in wavelength to enhance the detectability of absorption signals. (a) Typical geometric arrangement; (b) Wedged windows used in the laser pass.

Figure 6.7(a) shows typical TDLAS signals. The wavelength scanning methods are dependent on lasers used in the system. In DFB lasers, the laser wavelength is scanned by changing an input current to the laser and in ECDLs by changing the angle of mirrors or gratings inside the laser cavity. "Reference" and "signal" outputs are measured and the division of the two becomes the absorption spectra. "Reference" is the output without the path

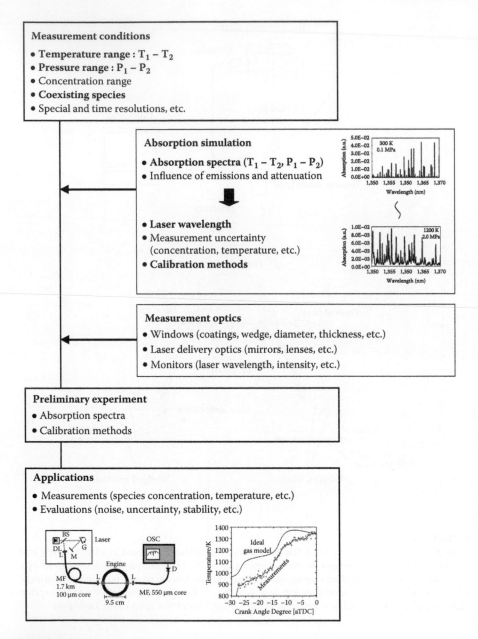

FIGURE 6.6

Application procedures in TDLAS. The most important factors for TDLAS are the selection of laser wavelength and the reduction of noise. Theoretical predictions of TDLAS spectra are important when selecting laser wavelength in practical applications.

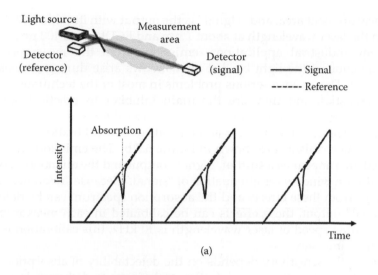

(a)

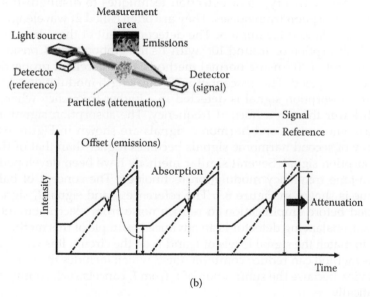

(b)

FIGURE 6.7
Typical TDLAS signals. "Reference" and "signal" outputs are measured, and the division of the two becomes the absorption spectra. In many industrial applications, emissions from the measured area and attenuation of the light by dirt on windows arise during a measurement period. TDLAS can cancel these effects, and this is an excellent feature of TDLAS. (a) "Reference" and "signal" outputs in a normal condition; (b) "Reference" and "signal" outputs with emissions and attenuation.

of the measurement area, and "signal" is the output with its path. DFB lasers can scan the laser wavelength at about 1 nm and ECDLs over 100 nm.

In many industrial applications, emissions from the measured area and attenuation of the light by dirt on windows arise during a measurement period. These cause serious problems in most of the techniques using laser diagnostics, and they are the main inhibitor to practical applications. However, TDLAS has the ability to cancel these effects, and this is an excellent feature of TDLAS. It is often called "self-calibration." The self-calibration mechanisms are shown in Figure 6.7(b). The emissions from the measured area appear as a shift of "signal" output and the attenuation of the light by dirt on windows as attenuation of "signal." Because "reference" does not change from these effects and the absorption spectrum can be detected as a "signal" output, these effects can be calibrated in each measurement. If the scanning speed of laser wavelength is 10 kHz, this calibration is also done at the same rate.

In TDLAS the sensitivity depends on the detectability of absorption, and there are mainly two types of detection techniques to distinguish a small amount of absorption from noises. They are often called 2f wavelength modulation and balanced techniques. The detection limit of these two methods can reach absorption of around 10^{-6} with 1 s integration time. Considering a detection limit of 10^{-3} using normal methods, their excellent properties can easily be understood. The concept of a 2f wavelength modulation technique is that the absorption signal is detected in a higher frequency where noise is much lower than in a normal frequency. The absorption signal and its first harmonic and second harmonic signals are shown in Figure 6.8. The frequency of second harmonic signals becomes higher than that of the normal absorption signal. Several similar methods have been developed, such as a two-tone frequency modulation technology. The concept of balanced technique is shown in Figure 6.9. The reference I_0 and signal I_λ signals are subtracted before amplification to reduce noise from electric circuits. The Hobbs auto-balancing detector is an advanced example of this method and it is used to match the signal levels of I_0 and I_λ on the circuit. It is worth noting that this method can reduce common noise, which includes the noise from a laser device, because the subtraction of I_0 from I_λ cancels such common noise automatically.

Because the line shape function is normalized as shown in Equation (6.8), the species number density is proportional to the area of the absorption spectrum. It is important to understand that the height of the peak in the absorption spectrum is not always directly proportional to the number density because the shape of spectra depends on broadening parameters. Broadening parameters are functions of temperature, pressure, and coexisting species concentrations. Figure 6.10 shows an effect of the line shape function to absorption spectra. There are three types of broadenings and only the collisional broadening FWHM is changed by 10% under the same

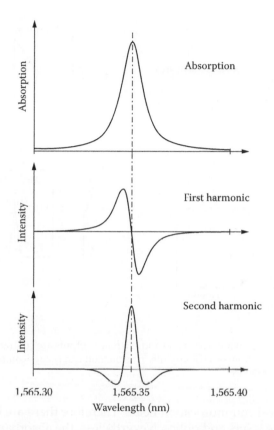

FIGURE 6.8
Absorption signal and its first harmonic and second harmonic signals. The concept of 2f wavelength modulation technique is that the absorption signal is detected in a higher frequency where noise is much lower than in a normal absorption signal. The frequency of second harmonic signals becomes higher than that of the normal absorption signal.

temperature and pressure conditions. This usually happens when the composition of coexisting species is changed by the mixing or reaction processes. Because the species number density is the same, the area of absorption is the same. However, the height of the peak changes by 10% in the absorption signal and by 30% in the 2f signal. If the heights of the direct or 2f spectra are used for a concentration measurement, deviations of the measured results arise depending on the measurement conditions. Care should be taken especially when handling the 2f wavelength modulation technique.

Molecules detected by TDLAS are shown in Table 6.2 with their absorption wavelength. These species include molecules such as H_2O, CO, NO, and so on. Many species have absorption lines at the near infrared (NIR, 0.7–2.5 µm) wavelength region, and these are often accessible using commercial DFB lasers. NIR lasers and optics are reasonably priced and robust because they

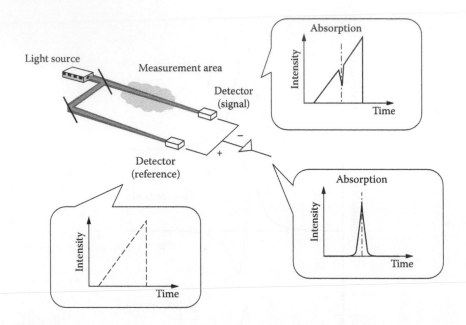

FIGURE 6.9
Concept of balanced technique. The signal and reference outputs are subtracted before amplification to reduce noise from electric circuits. This method can reduce common noise, including noise from a laser device.

are used in optical communication fields. Therefore there are lots of applications using these lasers and optics. Nevertheless, the absorbance in the mid-
infrared (MIR, 2.5–10 μm) wavelength region is usually much bigger than
that in NIR. This is because molecular overtone and combination absorption
bands are located at NIR wavelengths and fundamental bands are mainly
at MIRs. Figure 6.11 shows absorption spectra of NO and N_2O in 1.0–10 μm.
Absorption spectra of these molecules are calculated based on the HITRAN
database. It is clear that these molecules have bigger absorption around 5 μm.
In this wavelength region, other types of lasers, including quantum cascade
lasers, have to be employed. There are also challenging problems in MIR
for the delivery of laser light using fiber optics, because normal quartz optical fibers cannot transmit light within this wavelength region. Applications
using MIR lasers are still in progress.

TDLAS is often called a line-of-sight measurement technique and its measurement is based on the total amount of absorption along the laser path.
However, there is the technique called computer tomography (CT) that
reconstructs the two-dimensional (2-D) information by a set of absorption
signals. CT is widely used in the medical fields, and this technique has been
gradually applied to TDLAS. A set of laser paths goes through a measurement field, and the absorption signals are used to reconstruct the 2-D image

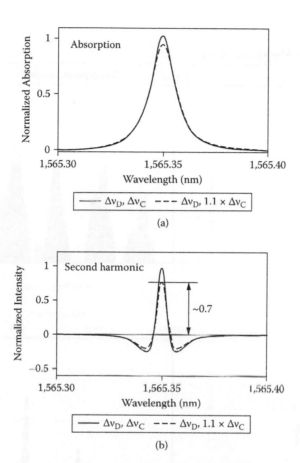

(a)

(b)

FIGURE 6.10
Effect of the line shape function to absorption spectra. The collisional broadening FWHM is changed by 10% under the same temperature and pressure conditions. Because the species number density is the same, the area of absorption is the same. However, the height of the peak changes by 10% in the absorption signal and by 30% in the 2f signal.

of a measured area as shown in Figure 6.12. The following equation can be given in each laser path:

$$A_j = \sum_{i=1} \alpha_i \ell_{ij} \tag{6.13}$$

where A_j is the integrated absorbance and ℓ_{ij} is the path length of the j path. Using a set of Equation (6.13), 2-D distributions of concentration and temperature are reconstructed by CT. One of the merits of TDLAS is its fast response; theoretically 2-D reconstruction can be done at a rate higher than kHz.

TABLE 6.2

Typical Species Absorption Spectra

Molecules	Absorption Spectra
Water vapor : H_2O	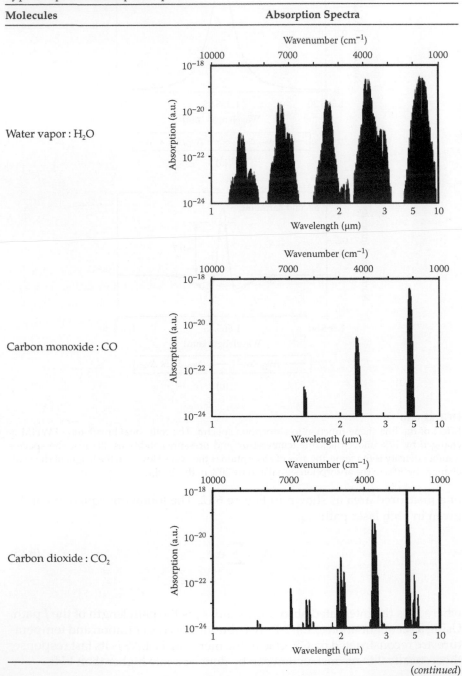
Carbon monoxide : CO	
Carbon dioxide : CO_2	

(continued)

TABLE 6.2 (CONTINUED)

Typical Species Absorption Spectra

Molecules	Absorption Spectra
Nitric oxide: NO	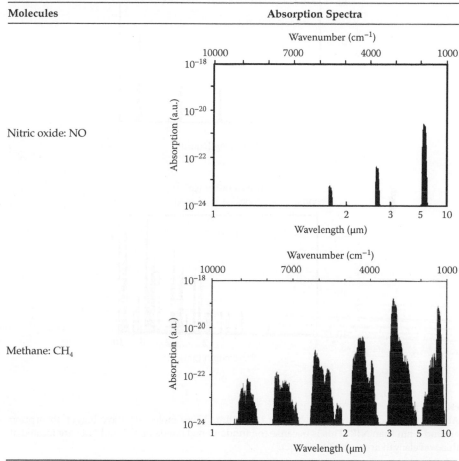
Methane: CH$_4$	

Note: Absorption spectra of H$_2$O, CO, CO$_2$, NO, CH$_4$, and so forth are synthesized using databases including HITRAN. It is worth noting that all the absorption lines are not always included in the databases; literature checks are also necessary to confirm the validity of the simulation results.

TDLAS can also be used for velocity measurement using the Doppler effect of light. This method can be applicable to the velocity measurements in the range of near or over the velocity of sound. Because of its reasonable cost and ruggedness, it has been used for mass flow monitoring. One of the setups of this method is shown in Figure 6.13.[6.4] Using two laser paths in opposite directions, the absorption spectrum of each path shows the Doppler shift at different wavelengths as shown in this figure.

There are also several methods that use almost the same principle as TDLAS. These include Fourier transform infrared spectroscopy (FTIR),

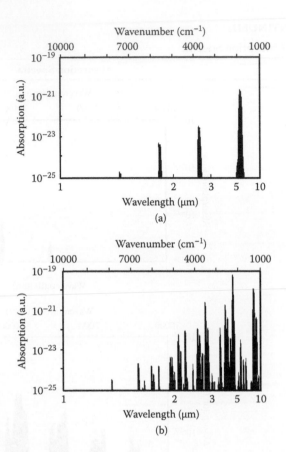

FIGURE 6.11
Absorption spectra of NO and N_2O in 1.0–10 μm. These molecules have bigger absorption around 5 μm than NIR. This is because the fundamental bands of NO and N_2O are located at this wavelength area. (a) NO; (b) N_2O.

nondispersive infrared spectroscopy (NDIR), photoacoustic spectroscopy (PAS), and cavity ring-down spectroscopy (CRDS).[6.5] PAS measures the effect of absorbed energy by atoms or molecules as an acoustic signal. The absorbed energy by atoms or molecules is transformed into heat by energy transfer processes, and this local heating process leads to a pressure wave of sound. PAS detects this photoacoustic signal instead of the decreased laser energy in TDLAS. CRDS is a sensitive technique that enables concentration measurements down to the ppt (parts per trillion) level. A typical setup of CRDS consists of two highly reflective mirrors, as shown in Figure 6.14. When the pulsed laser beam is introduced into the space between two mirrors, which is often called "cavity," the intensity of the laser exponentially decreases by leaking from the cavity. During this energy decrease, the laser beam is reflected back and forth numerous times between these two mirrors,

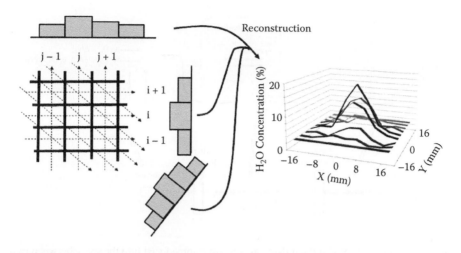

FIGURE 6.12
2-D image reconstruction by a set of absorption signals using CT. 2-D distributions of concentration and temperature are reconstructed by CT.

and this leads to a long path length on the order of a few kilometers. When atoms or molecules exist inside the cavity and these species have absorption lines in the scanned laser wavelength, the intensity of light decreases faster. Therefore this decaying time can be used to measure the concentration of atoms or molecules in the cavity. CRDS has excellent sensitivity and can be

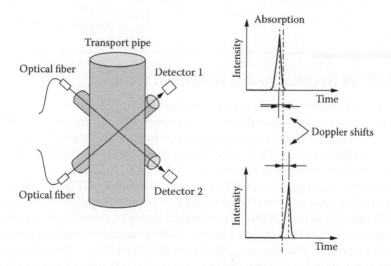

FIGURE 6.13
Typical setup of mass flow monitoring using TDLAS. Using two laser paths in opposite directions, the absorption spectrum of each path shows the Doppler shift at different wavelengths. Velocity is measured from the shift and concentration from the intensity of the absorption signal.

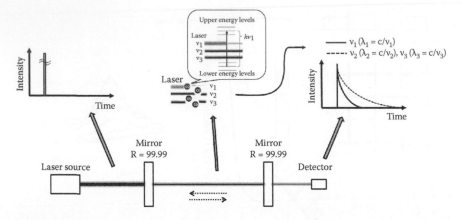

FIGURE 6.14

Typical setup of CRDS. When the pulsed laser beam is introduced into the space between two mirrors, the intensity of the laser exponentially decreases by leaking from the mirrors. During this energy decrease, the laser beam is reflected back and forth numerous times between these two mirrors, and this leads to a long path length on the order of a few kilometers. Absorption of atoms or molecules can be detected by the decay time of the laser.

applied to trace species detection, but it depends strongly on the reflectivity of two mirrors, which means contamination on mirrors is rather crucial in CRDS. It sometimes encounters difficulties in industrial applications, especially *in situ* measurements.

6.3 TDLAS Applications to Industrial Fields

TDLAS has been employed in many industrial applications,[6.6]–[6.20] including combustion and flow analyses, trace species detections, plasma processing, process monitoring and its control, environmental monitoring, and so on. TDLAS has been applied to both the clarification of basic phenomena in industrial processes and the monitoring and advanced control of industrial systems. The high sensitivity and fast response of TDLAS enable the monitoring of system control parameters in practical applications. The cost of a TDLAS unit is reasonable compared to those of other laser diagnostics, because tunable diode lasers and photo-diodes are much less expensive than Nd:YAG lasers and charge-coupled device (CCD) cameras. This is one of the biggest motivations to use this method for industrial applications. TDLAS is mainly used for gas measurements.

There are two types of approaches for TDLAS applications. One is the *in situ* measurement of species concentrations, temperature, pressure, and velocities. The laser beam is directly introduced into the measurement area,

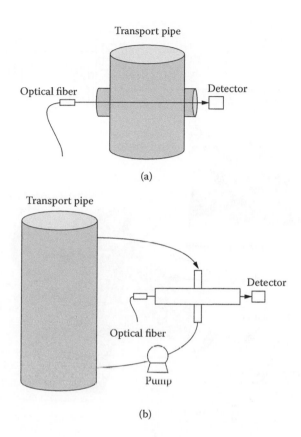

FIGURE 6.15
There are two types of approaches for TDLAS applications. One is the *in situ* measurement of species concentrations, temperature, pressure, and velocities. The laser beam is directly introduced into the measurement area. The other is based on extractive sampling, and gases are sampled and introduced into a measurement cell. (a) *In situ* measurement; (b) Extractive sampling.

as described in Figure 6.15(a). In these applications, fiber optics are often used to maintain the robustness and ease of use. The other approach is based on extractive sampling systems, shown in Figure 6.15(b). Gases are sampled and introduced into a measurement cell. A multipath cell, in which the laser beam is reflected using a set of mirrors to make a long path length, is often used to enhance the detectability.

6.3.1 Engine Applications

In car engines, an increasing concern in environmental issues—air pollution, global warming, and petroleum depletion, for example—has helped drive research. TDLAS has been applied to engine measurements in various ways,

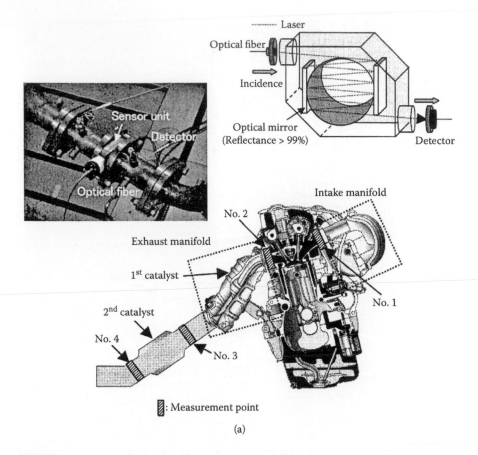

(a)

FIGURE 6.16
Application of TDLAS to engine exhausts and intake air measurements. The measurement response time is 1 ms to measure gas concentrations (CO, CO_2, H_2O, and CH_4) and temperature in each combustion cycle. A sensor was attached directly to a flange part of the piping. A clear difference in response is recognized between the measurements by TDLAS and a traditional sampling method. (a) Measurement positions and schematic of sensor unit; (b) Close-up of engine start; (c) Measurement results at intake manifold. (Reprinted from M. Yamakage, K. Muta, Y. Deguchi, S. Fukada, T. Iwase, and T. Yoshida, *SAE Paper* 20081298, 51, 2008. With permission from SAE International.)

which include intake air,[6.19] exhaust,[6.19] and engine cylinder measurements. Because of its fast response, TDLAS has a lot of merit in engine applications.

Figure 6.16 shows an application of TDLAS to engine exhaust gas and intake air measurements.[6.19] The measurement response time is 1 ms to measure gas concentrations (CO, CO_2, H_2O, and CH_4) and temperature in each combustion cycle. A sensor was attached directly to a flange part of the piping. The laser beam was guided to the sensor unit by an optical fiber and detected by a photo-diode after passing through the gas flow. In this sensor unit, mirrors set in parallel can reflect the laser beam 10 times and the

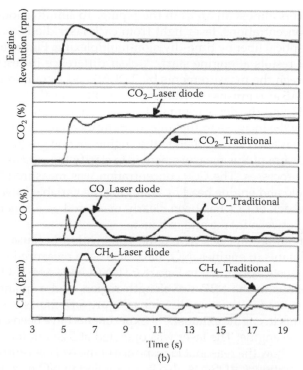

(b)

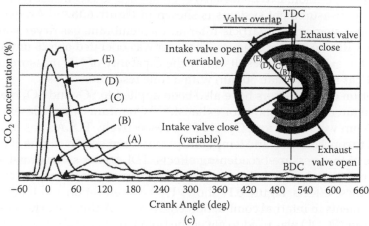

(c)

FIGURE 6.16
(Continued)

laser path covers almost all areas in the piping. To simultaneously measure several gas concentrations, laser light from each laser diode was combined into a single optical fiber by time-division-multiplexing. Figure 6.16(b) shows measurement results at engine start. A clear difference in response is recognized between the measurements by TDLAS and a traditional sampling method. TDLAS also grasps the transient phenomenon, that is, the increases in CO and CH_4 and the decrease in CO_2. These phenomena result from rich combustion by the increase in fuel injections immediately after engine start. This method has also been applied to the engine suction gas (intake air) measurement. The aim of this measurement is to clarify the exhaust gas recirculation (EGR) processes. The CO_2 concentration was measured between the intake manifold and engine head by changing the variable valve timing and lift. The blowback of CO_2 has been clearly measured as a function of valve overlap time between intake and exhaust.

NO$_x$ are important species in engine combustion because they are the main air pollutants in engines. NO$_x$ have also been measured using TDLAS in engine exhausts. As shown in Figure 6.11, NO, NO_2, and N_2O have strong absorption bands in the MIR wavelength region, and a quantum cascade laser is mainly used in these applications. Figure 6.17 shows *in situ* NO measurement results in engine exhaust.[6.21] Experiments have been carried out on a static gasoline engine. The laser was operated at 5.26 μm for the detection of NO. Results show the reasonable response of the NO concentration during the engine operation. CRDS has also been applied to NO measurement and one of the measurement systems is shown in Figure 6.18.[6.22] An extractive sampling system with a particle filter and a membrane gas dryer was used to apply CRDS to engine exhaust. The laser was operated at 5.26 μm to measure NO in exhaust. The results show the capability of NO measurement in exhaust gas for more than 30 min with a time resolution of 1 s.

Quantum cascade lasers have also been applied to SO_2 and SO_3 measurements in aircraft test combustor exhaust.[6.23] Quantum cascade lasers at 7.50 and 7.16 μm were used for SO_2 and SO_3, respectively. TDLAS has also been applied to concentration and temperature measurements in engine cylinders. Because of the pressure-broadening effects, TDLAS has some drawbacks in high-pressure fields, and their countermeasures are necessary to compensate for these effects. Figure 6.19 shows H_2O concentration and temperature measurements in internal combustion engines.[6.24] A tunable external-cavity diode laser (ECDL) was used to cover the broadened H_2O absorption spectra during combustion. The laser was scanned from 1374 to 1472 nm. Gas temperature and H_2O concentration were measured every 85 μs from each laser scan. An engine was operated in homogeneous-charge compression ignition (HCCI) mode. It was demonstrated that during the 35 crank-angle degrees of a single compression stroke, the measured temperature and H_2O mole fraction rose from 800 K and 0.3% to 1350 K and 2.7%, respectively.

There are other types of TDLAS applications where TDLAS is used as a sensor of gas temperature and concentration. Figure 6.20 shows a schematic

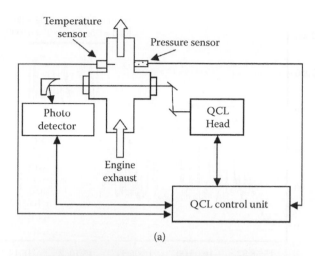

(a)

(b)

FIGURE 6.17

In situ NO measurement results in engine exhaust using TDLAS. A room-temperature continuous wave quantum cascade laser can be used for the *in situ* monitoring of vehicle exhausts. Nitric oxide concentrations have been determined in real time and clearly respond to changes in engine operating conditions. (a) Experimental setup for NO measurements on engine exhaust; (b) Vehicle engine test bed, NO measurements on engine exhaust; (c) NO concentration variation after the engine start. (Reprinted from V.L. Kasyutich, R.J. Holdsworth, and P.A. Martin, "*In situ* vehicle engine exhaust measurements of nitric oxide with a thermoelectrically cooled, cw DFB quantum cascade laser," *Journal of Physics: Conference Series* 157, 012006, 2009. With permission from the Institute of Physics.)

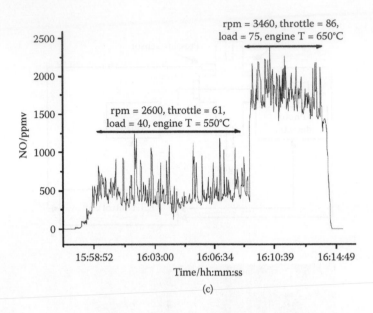

FIGURE 6.17
(Continued)

drawing of one of these applications.[6.25] The measurement device is embedded in a spark plug, and a 6-mm laser path next to the spark plug enables measurements of temperature and H_2O concentration near the spark plug. DFB lasers at 1345 and 1388 nm were employed with a 2f wavelength modulation technique. The temperature is determined from the absorption ratio of two transitions, and the H_2O concentration is determined from one of the absorption intensities using this inferred temperature. Using this configuration, temperature and H_2O concentration can be measured over a range of temperatures and pressures from 500 to 1050 K and 0.11 to 5 MPa at 7.5 kHz in internal combustion engines. Crank-angle-resolved temperature and H_2O concentration are shown in Figure 6.20(b) and 6.20(c), respectively.

The application of TDLAS with computer tomography has been also demonstrated in a multicylinder automotive engine. The merit of this method is the fast and continuous imaging of a 2-D measurement section, which is rather hard to attain by laser-induced fluorescence (LIF). The smaller size of the laser access ports is the other important feature of TDLAS. It is not necessary for TDLAS to need large access windows into combustion chambers, which is often necessary in LIF. Figure 6.21 shows an application of chemical species tomography in a multicylinder automotive engine.[6.26] Optical access ports, optical fibers, and collimators are embedded in the engine cylinder. A measurement grid consisting of 27 dual-wavelength optical paths has been implemented in one cylinder of a gasoline engine. Dual-wavelength optical paths have been employed to eliminate the attenuation of laser intensity

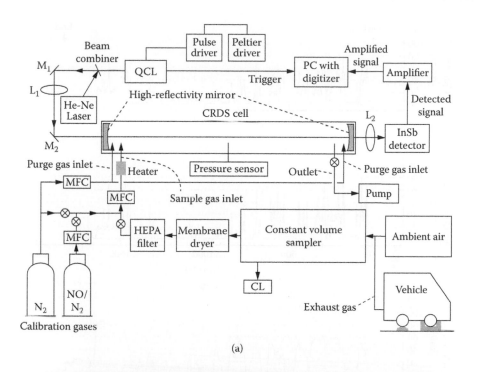

(a)

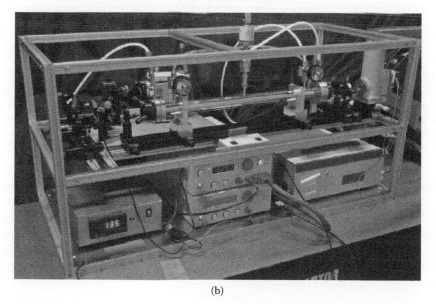

(b)

FIGURE 6.18a, b

NO measurement results in engine exhaust using CRDS. (a) Experimental setup of CRDS-based exhaust-gas NO sensor. (b) Photograph of the apparatus. The equipment was mounted on a 470 mm × 1040 mm × 500 mm rack for portability.

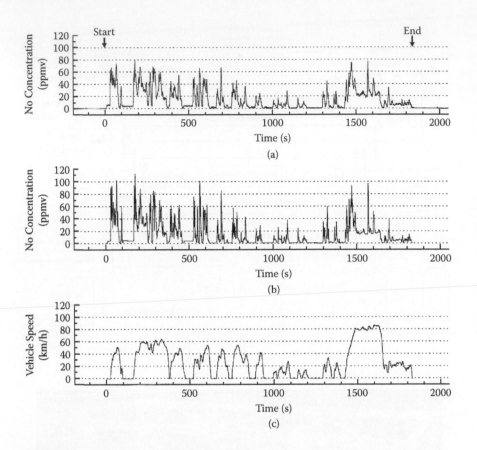

FIGURE 6.18c

NO measurement results in engine exhaust using CRDS. Real-time NO monitoring in the diluted exhaust gas by (a) CRDS and (b) conventional NO$_x$ sensor. (c) The vehicle speeds in the JE05 test. (Reprinted with kind permission from Springer Science+Business Media: *Applied Physics B*, 100(4), 925–931, "Real-time monitoring of nitric oxide in diesel exhaust gas by mid-infrared cavity ring-down spectroscopy," H. Sumizawa, H. Yamada, and K. Tonokura, 2010.)

other than the measured species (hydrocarbon fuels). The system employs a 1700-nm laser light for a hydrocarbon absorption measurement and a 1651-nm laser light unit as a reference. Laser intensity was modulated at frequencies up to 1 MHz and the effective frame rate was 3000–4000 frames per second. Measured tomographic images are shown in Figure 6.21(b). This method can detect rapid changes of fuel concentration distribution at a resolution of 3 degrees of crank angle.

6.3.2 Burner and Plant Applications

TDLAS has been applied to several types of burners, from a laboratory-scale burner to a large commercial-size burner. In laser diagnostics, many

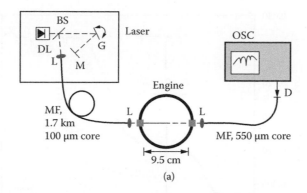

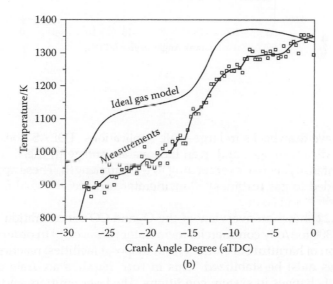

FIGURE 6.19

H_2O concentration and temperature measurements in engine cylinder. (a) Experimental setup including laser system, top view of combustion chamber, and detection system. *DL*, diode laser; *BS*, beam sampler; *G*, vibrating grating; *M*, metal mirror; *MF*, multimode fiber; *L*, lens; *D*, detector (photo-diode); and *OSC*, oscilloscope. (b) Measured and modeled gas temperature as a function of CAD. The measurement is made in a single engine cycle. The solid line within the measurement data corresponds to an average over three consecutive scans (0.92 CAD) and serves as guidance only. The model assumes a fixed number of moles in the engine; because of severe blow-by, the agreement is poor. However, the model does reveal the two-stage combustion process. (c) Measured and modeled water mole fraction as a function of CAD. The measurement is made in a single engine cycle. The solid line within the measurement data corresponds to an average over three consecutive scans (0.92 CAD) and serves as guidance only. The modeled curve is included for discussion purposes only and is not intended for direct comparison. The nonzero mole fraction before onset of combustion is assigned to recirculated H_2O produced in previous engine cycles. (Reprinted from *Proceedings of the Combustion Institute*, 30(1), L.A. Kranendonk, J.W. Walewski, T. Kim, and S.T. Sanders, "Wavelength-agile sensor applied for HCCI engine measurements," 1619–1627, Copyright 2005, with permission from Elsevier.)

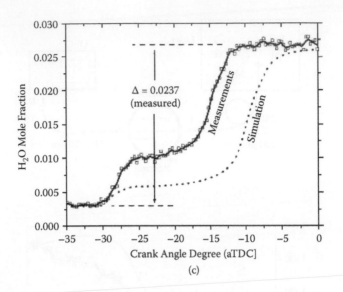

FIGURE 6.19
(Continued)

methods have drawbacks in large-scale applications. TDLAS does not have significant disadvantages and even has one advantage in these conditions: Its signal intensity increases according to the path length. These applications have extended to gas turbines,[6.27] incinerator furnaces,[6.16] coal-fired boiler burners,[6.28],[6.29] and so on.

Figure 6.22 shows an overview of the O_2 and CO concentration measurements in a 300 ton/day commercial incinerator furnace.[6.16] In order to reduce the emission of harmful substances from disposal facilities, reactions within the facilities must be stabilized. This in turn requires accurate and rapid detection of changes in system conditions. The laser emitters and receivers were each positioned on opposite surfaces of the furnace wall with a 9-m path length. CO measurement points were selected in the upper area of secondary air, with O_2 measurement at another location in the upper area. The existing conventional monitor was located at the bag filter outlet. In secondary air allocation control, the relative CO concentration measurement results are used as inputs for the control logic, and the allocation of the amount of secondary air (A and A′ side) is controlled based on the A and A′ CO concentration ratios.

Figure 6.22(b) indicates O_2 concentration trends for both laser measurements and at the bag filter outlet. The in-furnace laser measurements are capable of detecting O_2 and CO concentration 2–3 min faster than the existing conventional monitor. This 2–3 minute period is extremely valuable for control of combustion. Figure 6.22(c) shows CO concentration trends in normal operations. There are several CO peaks at both points A and A′. These CO peaks cannot be controlled by the normal control scheme using the existing O_2 and

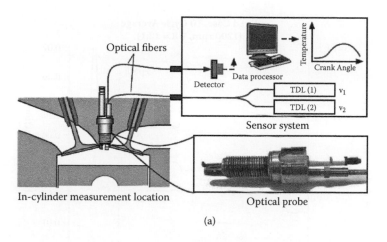

(a)

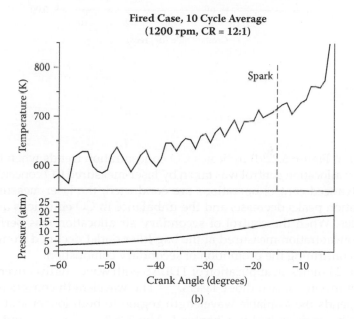

(b)

FIGURE 6.20

TDLAS sensor unit embedded in a spark plug. (a) Sensor layout; (b) Measured temperature and pressure in a fired, single-cylinder engine; (c) Water concentration and mole fraction measured in a fired, single-cylinder engine. (Reprinted from *Proceedings of the Combustion Institute*, 31(2), G.B. Rieker, H. Li, X. Liu, J.T.C. Liu, J.B. Jeffries, R.K. Hanson, M.G. Allen, S.D. Wehe, P.A. Mulhall, H.S. Kindle, A. Kakuho, K.R. Sholes, T. Matsuura, and S. Takatani, "Rapid measurements of temperature and H_2O concentration in IC engines with a spark plug-mounted diode laser sensor," 3041–3049, Copyright 2007, with permission from Elsevier.)

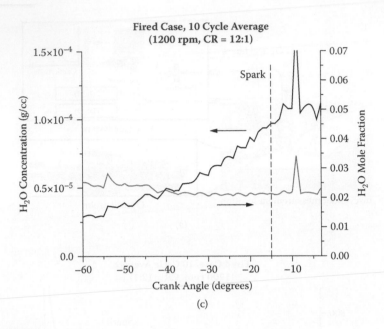

FIGURE 6.20
(Continued)

CO monitor. Figure 6.22(d) indicates CO concentration trends when the secondary air allocation control was taken by laser-measured CO concentration. When advanced control operations are conducted, the laser-measured CO concentration peaks decrease, and the imbalance in CO emissions at A and A′ subsides. When the control of secondary air allocation was performed, the CO concentration measured at the bag filter outlet declined from 11.9 to 8.0 ppm, confirming the promotion of secondary combustion.

Figure 6.23 shows an application of TDLAS with wavelength conversion by frequency mixing of two diode lasers.[6.28] The wavelength conversion technique expends the available wavelength region to both longer and shorter wavelengths as described in Chapter 1. The 226.8-nm laser output is produced by frequency mixing of a 395-nm external-cavity diode laser and a 532-nm laser using a beta-barium-borate crystal. It was used to measure NO in a 30 kW coal-fired boiler burner. The detection limit of 4.5 ppm-m/$\sqrt{Hz}$ at 700 K was demonstrated in coal-combustion exhaust at a maximum detection rate of 5 Hz. The same method has been applied for the detection of mercury.[6.24] A 254-nm beam is generated by frequency mixing of a 375-nm single-mode laser and a 784-nm DFB laser. The detection limits for *in situ* and extractive sampling were reported to be 0.3 and 0.1 ppb (parts per billion) over a 1-m path length, respectively.

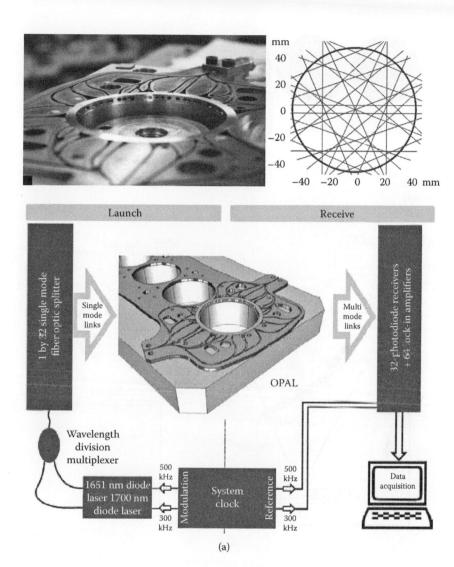

(a)

FIGURE 6.21

Application of chemical species tomography in a multicylinder automotive engine. (a) Schematic overview of the Imager system. The optical access layer (OPAL) during manufacture: fiber protection channels, coolant galleries, and a fiber entry/exit port (upper right) are all clearly visible. The 27-beam arrangement implemented by the present OPAL. (b) Five images showing rapid changes in the period from 42° to 30° before TDC at intervals of 3° of crank angle [1500 rpm/1.5 bar brake mean effective pressure (BMEP) load] obtained using the 15-beam subset. (Reprinted from *Chemical Engineering Journal*, 158(1), P. Wright, N. Terzija, J.L. Davidson, S. Garcia-Castillo, C. Garcia-Stewart, S. Pegrum, S. Colbourne, P. Turner, S.D. Crossley, T. Litt, S. Murray, K.B. Ozanyan, and H. McCann, "High-speed chemical species tomography in a multi-cylinder automotive engine," 2–10, 2010, with permission from Elsevier.)

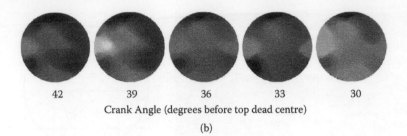

42   39   36   33   30

Crank Angle (degrees before top dead centre)

(b)

FIGURE 6.21
(Continued)

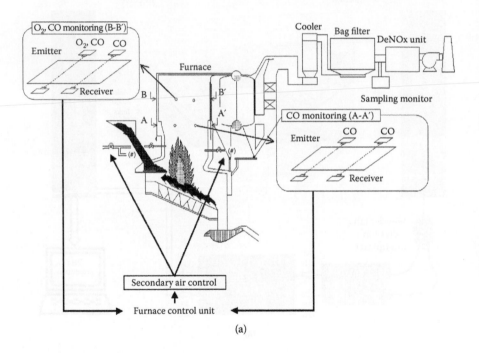

(a)

FIGURE 6.22

O_2 and CO concentration measurements in a 300 ton/day commercial incinerator furnace. The single in-furnace reference pass is used for O_2 measurement and the four passes for CO. The laser light for CO and O_2 measurements is mixed with a 680-nm guide laser light and separated inside the receiver box. Large amounts of furnace emissions were also reduced using narrow-band interference filters. The wavelength of the laser light was modulated at 2 kHz, and a total of 400 spectra were added to calculate each of the four pass concentrations. All functions are computer controlled, and the measured results are transferred to the incinerator control system at 2-s intervals. (a) O_2 and CO measurement points.; (b) Apparatus setup in the furnace; (c) O_2 measurement results; (d) CO emission results using existing control; (e) CO emission results using advanced control with a laser device. (Reprinted from Y. Deguchi, M. Noda, M. Abe, and M. Abe, "Improvement of combustion control through real-time measurement of O_2 and CO concentrations in incinerators using diode laser absorption spectroscopy," *Proceedings of the Combustion Institute*, 29(1), 147–153, 2002, with permission from the Institute of Physics.)

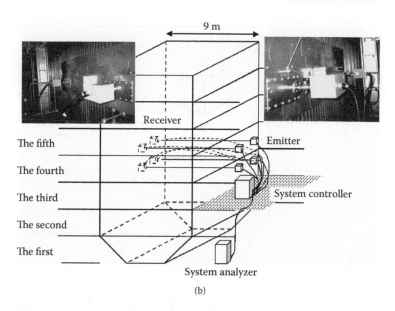

The fifth

The fourth

The third

The second

The first

Receiver

Emitter

System controller

System analyzer

(b)

FIGURE 6.22
(Continued)

6.3.3 Process Monitoring Applications

TDLAS is actively used for process monitoring because of its fast and noncontact features. Its reasonable cost enhances this trend. Applications cover the aluminum industry, steel-making industry, semiconductor industry, chemical industry, food and pharmaceutical industry, and so on. Currently, process monitoring in these industries mainly employs the conventional devices, and the key factors for TDLAS application are cost and reliability. The setup shown in Figure 6.6(b) is commonly used for *in situ* measurement. Some of the applications described earlier in the chapter also belong to this category.

There are several specific atoms and molecules that are useful in each industry. HF is an important species for aluminum-making industry because the aluminum smelting process uses alumina (Al_2O_3) and crinoline (Na_3AlF_6), resulting in the HF emission.[6.30] O_2, CO, and CO_2 are important species in many plants, including the steel-making industry[6.2],[6.30] and most combustion-related industries.[6.16] As described in Section 6.3.2, Figure 6.23 is the example of this application. NO_x are also important for the emission control from these processed. The same method shown in Figure 6.18 can be applicable in these industries.

There have been demonstrated several applications for chemical vapor deposition (CVD) process monitoring.[6.31],[6.32] CH_4 and C_2H_2 have been monitored using a quantum cascade laser at 7.84 nm.[6.31] A HCl measurement has also been demonstrated in a CVD process.[6.32] In the semiconductor industry, impurities such as H_2O affect plant performance. TDLAS has an excellent

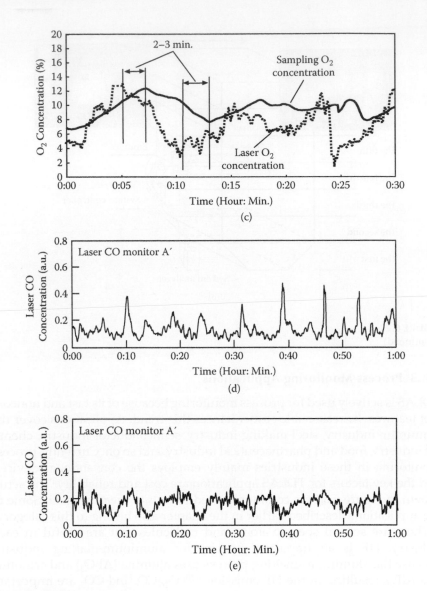

FIGURE 6.22
(Continued)

sensitivity for H_2O measurement. There has also been an application of H_2O mass flux monitoring and temperature in a freeze-drying process.[6.4],[6.33] The system is almost the same as that in Figure 6.14. The results have been used for a noncontact product temperature determination. The system using TDLAS is simple, and a similar system can be applicable over many industrial fields.

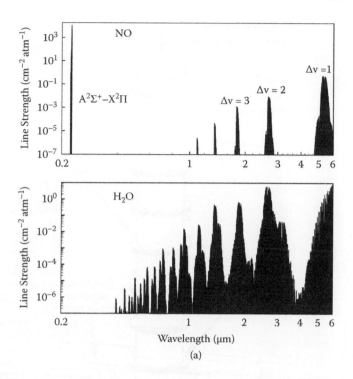

FIGURE 6.23
NO measurement in a 30-kW coal fired boiler burner using wavelength conversion by frequency mixing of two diode lasers. (a) Comparison of line strengths for NO and H_2O at 296 K. (b) Schematic of diode-laser-based sensor for NO measurements in coal-combustion exhaust. (c) Schematic of coal-fired boiler burner with reburn. (d) Summary of minimum NO mole fractions measured in the exhaust stream of the boiler burner with reburn for different fuel types and equivalence ratios. Mole fractions were measured *in situ* with the diode-laser sensor (TDLAS) and with extractive sampling and the EC sensor. Error bars represent the 10% uncertainty in diode-laser measurements. (Reprinted from T.N. Anderson et al., In situ measurements of nitric oxide in coal-combustion exhaust using a sensor based on a widely tunable external-cavity GaN diode laser," *Applied Optics*, 46(19), 3946–3957, 2007. With permission.)

6.4 Future Developments

Applications of TDLAS have extended to various industrial fields. In addition to the applications described in this chapter, it can be used for environmental monitoring, plant safety, and other purposes.[6.4] A detection of carbon isotopes of CO_2 has been also demonstrated in forest air monitoring. [6.34] It can also be employed as a miniature sensor[6.35] because the size of diode lasers is much smaller than other lasers, as shown in Chapter 1. It would not be an exaggeration to say that TDLAS at a NIR wavelength region is already a high-quality finished form for practical applications.

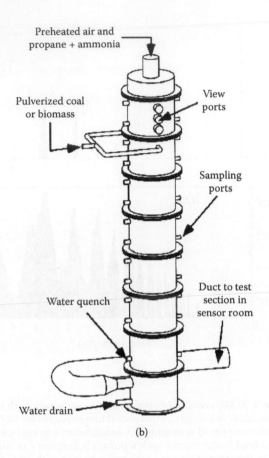

Preheated air and
propane + ammonia

Pulverized coal
or biomass

View
ports

Sampling
ports

Duct to test
section in
sensor room

Water quench

Water drain

(b)

FIGURE 6.23
(Continued)

However, in the MIR region there have been several challenges for the
advancement of TDLAS applications. The most notable and desired technology is the fiber delivery system in this wavelength region. Silica-based
fiber optics cannot be used in this wavelength region, and there is no reliable
and easy-to-use fiber delivery device. The optical fiber delivery system is an
important factor for industrial applications, and its development will make
TDLAS much more appealing to various industrial fields. The advancement
of lasers and detectors in MIR is also necessary to make the system reliable
and rugged. This technology also has a close relation to THz technology,
described in Chapter 8, because this wavelength region is located near the
THz frequency.

The other promising application of TDLAS is 2-D and 3-D measurements using CT. Currently, 2-D measurements fall in the category of LIF,
and TDLAS is considered as a line-of-sight technology. TDLAS with CT
has great potential for the fast and continuous imaging in a measured area,

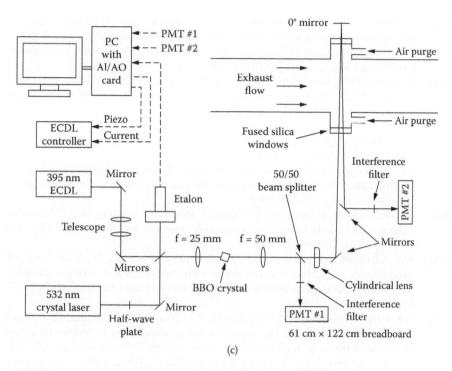

FIGURE 6.23
(Continued)

that cannot be attained by LIF. This technique is promising for both the clarification of basic phenomena in industrial processes and the monitoring and advanced control of industrial systems. The development of fast CT algorithms and CT optical systems is inevitable together with an advanced measurement technology.

References

[6.1]   L.S. Rothman, I.E. Gordon, A. Barbe, D.C. Benner, P.F. Bernath, M. Birk, V. Boudon, L.R. Brown, A. Campargue, J.-P. Champion, K. Chance, L.H. Coudert, V. Dana, M. Devi, S. Fally, J.-M. Flaud, R.R. Gamache, A. Goldman, D. Jacquemart, I. Kleiner, N. Lacome, W.J. Lafferty, J.-Y. Mandin, S.T. Massie, S.N. Mikhailenko, C.E. Miller, N. Moazzen-Ahmadir, O.V. Naumenko, A.V. Nikitin, J. Orphal, V.I. Perevalov, A. Perrini, A. Predoi-Cross, C.P. Rinsland, M. Rotger, F.M. Šimečková, M.A.H. Smith, K. Sung, S.A. Tashkun, J. Tennyson, R.A. Toth, A.C. Vandaele and J. Vander Auwera, "The HITRAN 2008 molecular spectroscopic database," *Journal of Quantitative Spectroscopy and Radiative Transfer*, 110(9–10), 533–572, 2009.

[6.2]   M. Lackner, "Tunable diode laser absorption spectroscopy (TDLAS) in the process industries-a review," *Reviews in Chemical Engineering*, 23(2), 5–147, 2007.

[6.3]   A.C. Eckbreth, *Laser Diagnostics for Combustion Temperature and Species*, Cambridge, Mass., ABACUS Press, 1988.

[6.4]   H. Gieseler, W.J. Kessler, M. Finson, S.J. Davis, P.A. Mulhall, V. Bons, D.J. Debo, and M.J. Pikal, "Evaluation of tunable diode laser absorption spectroscopy for in-process water vapor mass flux measurements during freeze drying," *Journal of Pharmaceutical Sciences*, 96(7), 1776–1793, 2007.

[6.5]   K. Kohse-Hoinghaus and J.B. Jeffries, *Applied Combustion Diagnostics*, New York: Taylor and Francis, 2002.

[6.6]   E.R. Furlong, D.S. Baer, and R.K. Hanson, "Real-time adaptive combustion control using diode-laser absorption sensors," *Symposium (International) on Combustion*, 27(1), 103–111, 1998.

[6.7]   G. Winnewisser, T. Drascher, T. Giesen, I. Pak, F. Schmulling, and R. Schieder, "The tunable diode laser: a versatile spectroscopic tool," *Spectrochimica Acta Part A*, 55(10), 2121–2142, 1999.

[6.8]   S.W. Allendorf, D.K. Ottesen, D.R. Hardesty, D. Goldstein, C.W. Smith, and A.P. Malcolmson, "Laser-based sensor for real-time measurement of offgas composition and temperature in BOF steelmaking," *Iron and Steel Engineer*, 75(4), 31–35, 1998.

[6.9]   P. Kohns, R. Stoermann, E. Budzynski, R.N. Walte, J. Knoop, and R. Kuester, "In-situ measurement of the water vapor concentration in industrial ovens by an user-friendly semiconductor laser system," *Proceedings of SPIE-The International Society for Optical Engineering*, 3098, 544–551, 1997.

[6.10] M.G. Allen, K.L. Carleton, S.J. Davis, W.J. Kessler, C.E. Otis, D.A. Palombo, and D.M. Sonnenfroh, "Ultrasensitive dual-beam absorption and gain spectroscopy: Applications for near-infrared and visible diode laser sensors," *Applied Optics*, 34(18), 3240–3249, 1995.

[6.11] D.B. Oh and D.C. Hovde, "Wavelength-modulation detection of acetylene with a near-infrared external-cavity diode laser," *Applied Optics*, 34(30), 7002–7005, 1995.

[6.12] D.M. Sonnenfroh and M.G. Allen, "Ultrasensitive, visible tunable diode laser detection of NO_2," *Applied Optics*, 35(21), 4053–4058, 1996.

[6.13] R.M. Mihalcea1, M.E. Webber, D.S. Baer, R.K. Hanson, G.S. Feller, and W.B. Chapman, "Diode-laser absorption measurements of CO_2, H_2O, N_2O, and NH_3 near 2:0 µm," *Applied Physics B*, 67(3), 283–288, 1998.

[6.14] S.J. Carey, H. McCann, F.P. Hindle, K.B. Ozanyan, D.E. Winterbone, and E. Clough, "Chemical species tomography by near infra-red absorption," *Chemical Engineering Journal*, 77(1–2), 111–118, 2000.

[6.15] V. Ebert, J. Fitzer, I. Gerstenberg, K.U. Pleban, H. Pitz, J. Wolfrum, M. Jochem, and J. Martin, "Simultaneous laser-based in situ detection of oxygen and water in a waste incinerator for active combustion control purposes," *Symposium (International) on Combustion*, 27(1), 1301–1308, 1998.

[6.16] Y. Deguchi, M. Noda, M. Abe, and M. Abe, "Improvement of combustion control through real-time measurement of O_2 and CO concentrations in incinerators using diode laser absorption spectroscopy," *Proceedings of the Combustion Institute*, 29(1), 147–153, 2002.

[6.17] S. Barrass , Y. Gérard , R.J. Holdsworth , and P.A. Martin, "Near-infrared tunable diode laser spectrometer for the remote sensing of vehicle emissions," *Spectrochimica Acta Part A*, 60(14), 3353–3360, 2004.

[6.18] M. Lewander, Z.G. Guan, L. Persson, A. Olsson, and S. Svanberg, "Food monitoring based on diode laser gas spectroscopy," *Applied Physics B*, 93(2–3), 619–625, 2008.

[6.19] M. Yamakage, K. Muta, Y. Deguchi, S. Fukada, T. Iwase, and T. Yoshida, "Development of direct and fast response gas measurement," *SAE Paper* 20081298, 51–59, 2008.

[6.20] C. Wang and P. Sahay, "Breath analysis using laser spectroscopic techniques: Breath biomarkers, spectral fingerprints, and detection limits," *Sensors*, 9, 8230–8262, 2009.

[6.21] V.L. Kasyutich, R.J. Holdsworth, and P.A. Martin, "In situ vehicle engine exhaust measurements of nitric oxide with a thermoelectrically cooled, cw DFB quantum cascade laser," *Journal of Physics: Conference Series* 157, 012006, 2009.

[6.22] H. Sumizawa, H. Yamada, and K. Tonokura, "Real-time monitoring of nitric oxide in diesel exhaust gas by mid-infrared cavity ring-down spectroscopy," *Applied Physics B*, 100(4), 925–931, 2010.

[6.23] W.T. Rawlins, J.M. Hensley, D.M. Sonnenfroh, D.B. Oakes, and M.G. Allen, "Quantum cascade laser sensor for SO_2 and SO_3 for application to combustor exhaust streams," *Applied Optics*, 44(31), 6635–6643, 2005.

[6.24] L.A. Kranendonk, J.W. Walewski, T. Kim, and S.T. Sanders, "Wavelength-agile sensor applied for HCCI engine measurements," *Proceedings of the Combustion Institute*, 30(1), 1619–1627, 2005.

[6.25] G.B. Rieker, H. Li, X. Liu, J.T.C. Liu, J.B. Jeffries, R.K. Hanson, M.G. Allen, S.D. Wehe, P.A. Mulhall, H.S. Kindle, A. Kakuho, K.R. Sholes, T. Matsuura, and S. Takatani, "Rapid measurements of temperature and H_2O concentration in IC engines with a spark plug-mounted diode laser sensor," *Proceedings of the Combustion Institute*, 31 (2), 3041–3049, 2007.

[6.26] P. Wright, N. Terzija, J.L. Davidson, S. Garcia-Castillo, C. Garcia-Stewart, S. Pegrum, S. Colbourne, P. Turner, S.D. Crossley, T. Litt, S. Murray, K.B. Ozanyan, and H. McCann, "High-speed chemical species tomography in a multi-cylinder automotive engine," *Chemical Engineering Journal*, 158(1), 2–10, 2010.

[6.27] X. Liu, J.B. Jeffries, R.K. Hanson, K.M. Hinckley, M.A. Woodmansee, "Development of a tunable diode laser sensor for measurements of gas turbine exhaust temperature," *Applied Physics B*, 82(3), 469–478, 2006.

[6.28] T.N. Anderson, R.P. Lucht, S. Priyadarsan, K. Annamalai, and J.A. Caton, "In situ measurements of nitric oxide in coal-combustion exhaust using a sensor based on a widely tunable external-cavity GaN diode laser," *Applied Optics*, 46(19), 3946–3957, 2007.

[6.29] J.K. Magnuson, T. N. Anderson, and R.P. Lucht, "Application of a diode-laser-based ultraviolet absorption sensor for in situ measurements of atomic mercury in coal-combustion exhaust," *Energy and Fuels*, 22(5), 3029–3036, 2008.

[6.30] I. Linnerud, P. Kaspersen, and T. Jæger, "Gas monitoring in the process industry using diode laser spectroscopy," *Applied Physics B: Lasers and Optics*, 67(3), 297–305, 1998.

[6.31] J. Ma, A. Cheesman, M.N.R. Ashfold, K.G. Hay, S. Wright, N. Langford, G. Duxbury, and Y.A. Mankelevich, "Quantum cascade laser investigations of CH_4 and C_2H_2 interconversion in hydrocarbon/H_2 gas mixtures during microwave plasma enhanced chemical vapor deposition of diamond," *Journal of Applied Physics*, 106(3), 033305/1–033305/15, 2009.

[6.32] V. Hopfe, D.W. Sheel, C.I.M.A. Spee, R. Tell, P. Martin, A. Beil, M. Pemble, R. Weissi, U. Vogth, and W. Graehlerta, "In-situ monitoring for CVD processes," *Thin Solid Films*, 442(1,2), 60–65, 2003.

[6.33] S.C. Schneid, H. Gieseler, W.J. Kessler, and M.J. Pikal, "Non-invasive product temperature determination primary drying using tunable diode laser absorption spectroscopy," *Journal of Pharmaceutical Sciences*, 98(9), 3406–3418, 2009.

[6.34] S.M. Schaeffer, J.B. Miller, B.H. Vaughn, J.W.C. White, and D.R. Bowling, "Long-term field performance of a tunable diode laser absorption spectrometer for analysis of carbon isotopes of CO_2 in forest air," *Atmospheric Chemistry and Physics*, 8(17), 5263–5277, 2008.

[6.35] M.B. Frish, R.T. Wainner, M.C. Laderer, B.D. Green, and M.G. Allen, "Standoff and miniature chemical vapor detectors based on tunable diode laser absorption spectroscopy," IEEE Sensors Journal, 10(3), 639–646, 2010.

7

Laser Ionization Time-of-Flight Mass Spectrometry

7.1 Principle

The principle behind laser ionization time-of-flight mass spectrometry (LI-TOFMS) is illustrated in Figure 7.1. A measurement sample is introduced into a vacuum chamber, and it is ionized by laser irradiation. The electric field potential V is simultaneously applied for acceleration of ions. The accelerated ions enter a region with no potential difference (the drift region) and undergo uniform motion. At this time, due to the law of energy conservation, the ions' electric field potential is equivalent to their kinetic energy. The following formula is established by the energy conservation low:[7.1]

$$zV = \frac{1}{2}m\left(\frac{\ell}{t}\right)^2 \tag{7.1}$$

where z is the ion value, ℓ the length of the drift region, m the ion mass, and t the time of flight of the ion traversing the drift region.

From Equation (7.1),

$$m = \frac{2zV}{\ell^2}t^2 \tag{7.2}$$

can be obtained, whereby the molecular mass of the sample is proportional to the square of the time of flight. Therefore the ion mass can determine the time of flight t required for the ion to reach the detector.

In laser ionizations, atoms or molecules are ionized through single or multiple photon (multiphoton) ionization processes. Especially, the resonance-enhanced two-photon ionization (often called REMPI) has been intensively applied to detect aromatic hydrocarbons and their derivatives. When light with high energy density (such as laser light) is introduced to molecules, the excited molecules then absorb this light as well and become ionized. This process is harnessed to ionize atoms or molecules, and then time-of-flight mass spectrometry is used to separate and detect targets from impurities. Energy

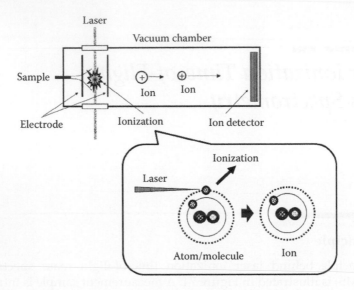

FIGURE 7.1
Principle behind LI-TOFMS. Unlike other laser diagnostics, a laser ionization TOFMS method samples the gas to a vacuum chamber. In the TOFMS process, ions can be counted by the ion detector; super-high sensitivity can be achieved using this method.

transfer processes of single-photon and multiphoton ionization are shown in Figure 7.2. These processes are simple and easy to understand. The merit of single-photon ionization is the large ionization cross section; its efficiency is much bigger than that of multiphoton ionization. In single-photon ionization, photofragmentation (the dissociation of excited molecules by incident laser light) is also small compared to multiphoton ionization. Because of this feature, single-photon ionization is often called as a "soft" ionization technique. The excitation wavelength of single-photon ionization often lies in a vacuum ultraviolet wavelength region, and this usually makes the measurement system much more complicated than that of multiphoton ionization.

Resonance-enhanced multiphoton ionization can cover the shortcomings of single-photon ionization using resonance states as steps to the ionization energy levels. The transfer process of resonance-enhanced two-photon ionization is shown in Figure 7.3 and can be written by the following rate equations:[7.2]

$$\frac{dn_1}{dt} = -n_1 W_{12} + n_2 (W_{21} + A_{21}) \tag{7.3}$$

$$\frac{dn_2}{dt} = n_1 W_{12} - n_2 (W_{21} + A_{21} + D_2 + W_I) \tag{7.4}$$

$$\frac{dn_i}{dt} = n_2 W_I \tag{7.5}$$

One-photon Ionization

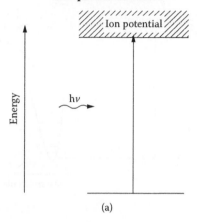

(a)

Multiphoton Ionization

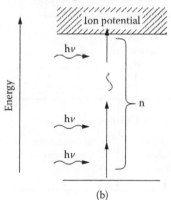

(b)

FIGURE 7.2
Energy transfer processes of single-photon and multiphoton ionization. The merit of single-photon ionization is the large ionization cross section; its efficiency is much bigger than that of multiphoton ionization. The excitation wavelength of single-photon ionization often lies at a vacuum ultraviolet wavelength region, and this usually makes the measurement system much more complicated than that of multiphoton ionization. (a) One-photon ionization; (b) Multiphoton ionization.

where n_1, n_2, and n_i are the number densities in the 1, 2, and ionization states respectively, A the Einstein A coefficient, W the stimulated emission or absorption rate, and D_2 the dissociation rate. W_{12}, W_{21}, W_I are the stimulated emission, absorption, and ionization rates, respectively, and they are proportional to the incident laser light intensity. Although a normal two-photon ionization rate depends on the square of the laser energy, the ionization rate of REMPI generally depends on all of these parameters. In the case of the

Resonance Enhanced Two-photon Ionization

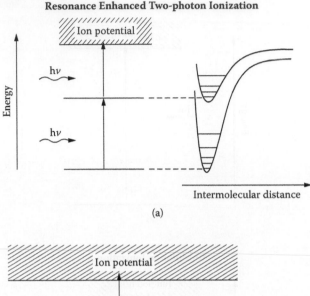

(a)

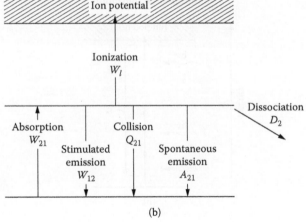

(b)

FIGURE 7.3
Energy transfer processes of REMPI. Resonance-enhanced multiphoton ionization can cover the shortcomings of single-photon ionization using resonance states as steps to the ionization energy levels. Enhancement of detection limits has been achieved in various methods, including pulsed ultrasonic jet and ion trap techniques. (a) Energy transfer process; (b) Laser ionization model.

large ionization cross section—that is, $W_I \gg W_{21} + A_{21} + Q_{21} + D_2$—the ionization rate depends largely on the excitation rate from "1" energy state to "2," that is, W_{21}. This means that the REMPI ionization rate is dependent on the parameters described above and on experimental conditions.

Enhancement of detection limits has been achieved in various methods, including pulsed ultrasonic jet and ion trap techniques, which have been used in various industrial applications. In the pulsed ultrasonic jet technique, the measurement of molecules is made thorough cryogenic cooling

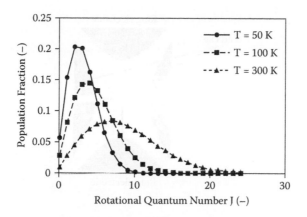

FIGURE 7.4

Boltzmann distribution at different temperatures. In the pulsed ultrasonic jet cooling process, the temperature becomes very low, e.g., a few tens of Kelvin. As the temperature becomes lower, molecules exist only at energy levels with small rotational quantum numbers. The rotational constant $B = 2$ cm^{-1} and vibrational constant $\omega_e = 2300$ cm^{-1} are used to evaluate the Boltzmann distributions.

by the ultrasonic jet. Cooled molecules are mainly located at several energy levels by the Boltzmann equation, and they are efficiently ionized by the resonance-enhanced ionization wavelength, thereby achieving improved selectivity and high sensitivity. In this method, tunable lasers are reused for tuning the laser wavelength to the resonance frequency of molecules. The number density of the molecule at a (v,J) ro-vibrational energy level is given by the Boltzmann equation

$$n_{v,j} = n \frac{g_{v,j} e^{-E_{v,j}/kT}}{\sum\limits_{v,j} g_{v,j} e^{-E_{v',j'}/kT}} \tag{7.6}$$

where $g_{v,J}$ and $E_{v,J}$ are the degeneracy and the energy of the (v,J) level, respectively (see Appendix D). Boltzmann distributions at different temperatures calculated by Equation (7.6) are illustrated in Figure 7.4. In the pulsed ultrasonic jet cooling process, the temperature becomes very low, a few several tens of Kelvin, and it is usually enough consider the rotational energy levels. As the temperature becomes lower, the molecules exist only at energy levels with small rotational quantum numbers. The number density fraction at low energy levels also increases at low temperature. The former feature leads to the better selectivity and the latter to the enhancement of sensitivity.

On the other hand, the ion trap system functions as a device for selecting mass, and it is capable of storing ions with certain mass numbers using the electric field inside the trap. It is also possible to distinguish ions with the

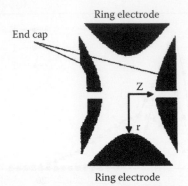

FIGURE 7.5
Structure of the ion trap ring electrode. The ion trap system functions as a device for selecting mass, and it is also capable of storing ions with certain mass numbers using the electric
field inside the trap. It operates by applying AC and DC electric fields to a hyperboloid-shaped
electrode.

same mass number using the tandem mass spectrometry (MS/MS) technology. Ions are introduced into an ion trap, which operates by applying AC and
DC electric fields to a hyperboloid-shaped electrode as shown in Figure 7.5.
When DC voltage U and AC voltage $V \cos wt$ are applied within this electrode, the electric potential Φ at point (r,z,t), describing cylindrical coordi-
nates within the trap, can be expressed by Equation (7.7)

$$\Phi(r, z, t) = \frac{\Phi_0(t)\ (r^2 - 2z^2)}{2r_0^2} + \frac{\Phi_0(t)}{2} \tag{7.7}$$

Here, for $\Phi_0(t) = U + V \cos wt$, r_0 is the minimum radius of the cell, or $2z_0^2 = r_0^2$. In this electric field, the movement of a charged particle with mass m and
electrical charge e becomes the same in both the r and z directions, and can
be expressed according to the Mathieu differential equations

$$\frac{d^2r}{dt^2} + \left(\frac{e}{mr_0^2}\right)(U + V \cos wt)r = 0 \tag{7.8}$$

$$\frac{d^2z}{dt^2} - \left(\frac{2e}{mr_0^2}\right)(U + V \cos wt)z = 0 \tag{7.9}$$

The trapped ions stay in a limited domain represented by the stable solutions
to Equations (7.8) and (7.9). Since the cell within the ion trap apparatus is of
a certain size (minimum radius r_0 of the hyperboloid), the trapped ion mass
depends upon the frequency of the AC voltage. Accordingly, species selectivity can be assigned to the ions to be trapped.

TABLE 7.1

Characteristics of LI-TOFMS

	Characteristics	Countermeasure
Theoretical treatments	• Rate equations	
Temperature	• Boltzmann distribution (lower energy states)	• Theoretical evaluation
Pressure	(Gas is sampled to a vacuum chamber)	
Windows	• Sensitive	• Heating
Calibration	• Easy	• Internal calibration
Noise	• Fragmentation • Memory effects	• Adjustment of laser wavelength and power • Heating
Measurement item	• Concentration	
Measurement dimension	• Point	
Detection limit	• ppt–ppb	
Stability	• Chamber (contamination) • Sampling pipes (contamination) • Laser	• Heating

Note: Fragmentation and memory effects are the important factors for the sensitive measurements. Fragmentation often causes noise and low signal intensity. Memory effects are caused by the adsorption and desorption processes of molecules on the surface of a sampling pipe. Adsorbed molecules on a metal surface tend to be desorbed for a prolonged period of time and the signals of desorbed molecules are continuously detected as noise.

There are some important phenomena involved in practical applications of this method. They are briefly summarized in Table 7.1. Fragmentation and memory effects are the important factors for the sensitive measurements.

1. *Fragmentation.* Though molecules can be ionized by single- or multiphoton ionization processes, at the same time, they also make fragments of molecules through a dissociation process. Fragmentation often causes noise and low signal intensity. There are several techniques for reducing the fragmentation process, described below.

2. *Contamination of a flight chamber.* Contamination of a flight chamber (including laser input windows) is crucial for the sensitive measurement. Because TOFMS is used for the detection of trace species, desorption of adsorbed species into the ionization area can cause significant noise, and it tends to last for a long period of time. This effect is almost the same as memory effects; heating of a flight chamber is one of the precautions to reduce the contamination of the chamber.

3. *Memory effects.* Memory effects are caused by the adsorption and desorption processes of molecules on the surface of a sampling pipe. Adsorbed molecules on a metal surface tend to be desorbed for a

prolonged period of time and the signals of desorbed molecules are continuously detected as noise even if there is no such component in the measurement (sampled) area. In LI-TOFMS, the sampling system is the essential device, and memory effects are inevitable in this technique. The surface condition affects the desorption rate; electropolished stainless steel pipes are often used as a sampling pipe to reduce these effects. Heating of the pipe also reduces memory effects.

4. *Stability of laser devices.* When LI-TOFMS is applied to monitoring in plants, the stability of the laser becomes a critical factor to maintain the stable operation of the measurement system. Several laser systems have been used for LI-TOFMS; however, these lasers are not rugged enough for long-term operation, such as two years with minimal maintenance. Therefore careful selection of lasers and a maintenance plan is necessary to apply this technique to industries.

7.2 Geometric Arrangement and Measurement Species

A typical geometric arrangement of laser ionization TOFMS is shown in Figure 7.6. In this method gases are sampled into a vacuum chamber for mass spectrometry. Pulsed lasers are used as an ionization source, and laser light is irradiated to the sampled gas in the vacuum chamber. Ionized atoms or molecules are accelerated and enter into a flight tube. They are detected by ion detectors including microchannel plates (MCPs). According to Equation (7.2), the mass of ion is proportional to the square of flight time, so the mass resolution becomes worse at a higher mass region. Because of the relation of flight time and flight length (length of the drift region), it is necessary to use longer flight tubes to maintain the mass resolution. A reflectron is often used to make a system compact and to have better mass resolution. It uses a static electric field at the end of the flight tube to direct ions to the opposite direction, as shown in Figure 7.6(b).

In practical applications, an internal calibration gas is often used to ensure a quantitative measurement. A certain amount of internal calibration gas is added to the sampling gas, and the measured ion signals are normalized by that of the internal calibration gas. A specific molecule that does not exist in the measurement field and has a similar ionization feature as that of target molecules is usually selected as an internal calibration gas. These include deuterated molecules such as C_6D_6. This calibration method can cancel or eliminate several unfavorable effects, such as fluctuation of laser intensity, contamination of measurement windows, and a malfunction of the TOFMS system.

When LI-TOFMS is used for the practical applications, users have to consider the factors described in Section 7.1. Figure 7.7 illustrates experimental

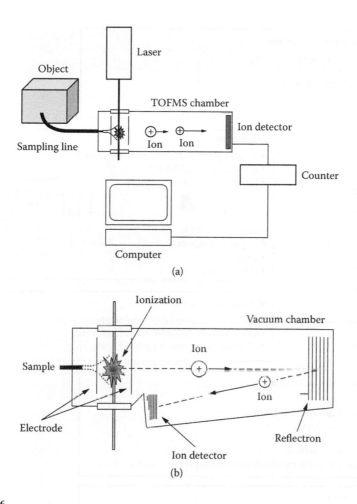

FIGURE 7.6

Typical geometric arrangement of LI-TOFMS. The main components are a laser and a TOFMS chamber. Supersonic jet cooling and ion trap methods are often used to enhance the sensitivity. A reflectron is often used to make a system compact and to have better mass resolution. (a) Typical geometric arrangement of LI-TOFMS; (b) Reflectron-type TOFMS.

procedures that have to be considered in LI-TOFMS applications. The most important factor for TOFMS is appropriate sampling conditions. TOFMS is a very sensitive device and vulnerable to memory effects. Sampling of materials with a high-boiling temperature at high concentrations will induce serious contamination in the sampling tubes and test and flight chambers. Clogging of an ultrasonic jet valve is a typical example of these troubles. Therefore sampling tubes and TOFMS chambers are often heated to reduce these effects. The LI-TOFMS timing chart is illustrated in Figure 7.8. The mass spectrum appears with a time delay from the laser ionization. Ion signals are

Measurement Conditions

- Temperature range : $T_1 - T_2$
- Pressure range : $P_1 - P_2$
- **Concentration range**
- Coexisting species
- Special and time resolutions, etc.

TOFMS simulation

- Excitation spectra (resonance wavelength)
- Ion trajectory in TOFMS
- Influence of coexisting species

- Laser wavelength
- Applied voltage in TOFMS
- Time resolution
- Calibration method

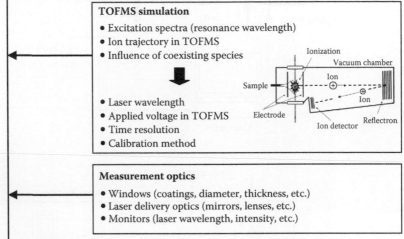

Measurement optics

- Windows (coatings, diameter, thickness, etc.)
- Laser delivery optics (mirrors, lenses, etc.)
- Monitors (laser wavelength, intensity, etc.)

Preliminary experiment

- Fragmentation
- Applied voltage in TOFMS
- Excitation spectra
- Mass spectra
- Calibration method
- Signal stability at measurement conditions $(T_1 - T_2, P_1 - P_2)$

Application

- Measurements (species concentration)
- Evaluations (noise, uncertainty, stability, etc.)

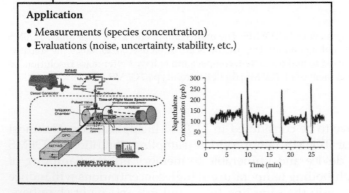

FIGURE 7.7

Application procedures in LI-TOFMS. The most important factor for TOFMS is appropriate sampling conditions. TOFMS is a really sensitive device and vulnerable to memory effects. Sampling of materials with high temperatures at high concentrations will induce serious contamination in the sampling tubes and test and flight chambers.

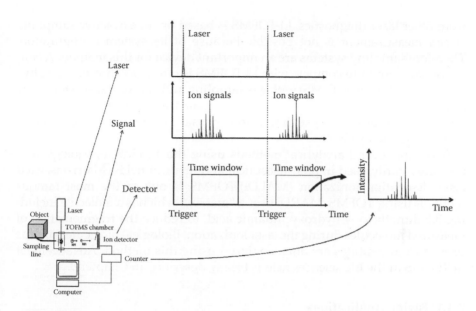

FIGURE 7.8

LI-TOFMS timing chart. The mass spectrum appears with a time delay from the laser ionization. Ion signals are measured by counting the ion signal from the detector within an appropriate time window.

measured by counting the ion signals from the detector within an appropriate time window.

Molecules detected by LI-TOFMS have been polycyclic aromatic hydrocarbons (PAHs), dioxins (DXNs), and polychlorinated biphenyls (PCBs). Their ionization wavelengths are in the UV range (240–300 nm) in the case of REMPI. Many species can be detected by LI-TOFMS; however, this method is usually used to detect molecules such as PAHs to which other methods are not accessible because of their low concentration. It is not a good choice to choose LI-TOFMS for a detection of H_2O in % ppm. Since H_2O can be detected by methods such as spontaneous Raman spectroscopy, TDLAS, or a conventional hygrometer, it is not reasonable to apply an expensive LI-TOFMS system for this purpose. If all of these methods are applicable, a conventional hygrometer will be the best choice. Therefore the concentration of the measured species using LI-TOFMS is often ppt or less.

7.3 Laser Ionization TOFMS Applications to Industrial Fields

LI-TOFMS has been employed to measure trace species at the range of ppb–ppt.[7.3]–[7.7] Though LI-TOFMS can be applied to molecules with small mass like NO, it is usually utilized to detect molecules with higher mass. Different from other laser diagnostics, LI-TOFMS is based on an extractive sampling; *in situ* measurement is not possible because of its system configuration. Therefore sampling systems are an important device for this method. A conventional method to compete with LI-TOFMS is often gas chromatography–mass spectrometry (GC-MS) with gas sampling and concentration steps. The advantage of LI-TOFMS is a fast response compared to the GC-MS method, and a few minutes' response time is often possible compared to a few weeks' analysis period by GC-MS.

There are several analytical methods using the TOFMS technology, and they are mainly used in pharmaceutical and medical fields. Matrix-assisted laser desorption/ionization (MALDI)-TOFMS is one of the most famous methods using TOFMS. MARDI uses a "matrix," which is a molecule including 3,5-dimethoxy-4-hydroxycinnamic acid, to reduce the fragmentation of measured molecules during the laser ionization. Biological molecules such as proteins and peptides are often analyzed using this method. Advancement in TOFMS in the life sciences field is briefly described in Chapter 8.

7.3.1 Engine Applications

In engine exhaust there are several toxic emissions including NO_x, CO, soot (or particulate), and polycyclic aromatic hydrocarbons (PAHs). They have been detected using laser diagnostics such as LIF, TDLAS, and laser-induced incandescence (LII). LIF can also be applicable to the detection of PAHs; however, the detection limit is not enough and species selective measurement is not possible because the fluorescence characteristics of PAHs are similar to each other. LI-TOFMS can detect PAHs individually with sufficient sensitivity.[7.8]–[7.10] Although LI-TOFMS can also detect NO_x and CO, it is rare to apply LI-TOFMS for the detection of these small molecules because other methods are much more suitable than LI-TOFMS.

Figure 7.9 shows an application of jet resonance-enhanced multiphoton ionization (REMPI) TOFMS to measure emissions of mono- and polycyclic aromatic hydrocarbons from a 60-kW diesel generator during steady state and transient operations.[7.8] The Nd:YAG pumped optical parametric oscillator (OPO) is used as a light source to tune the laser wavelength to maximize the detection sensitivity for each molecule, for example, 259 nm for benzene. PAHs were detected in the concentration range of ppb in the diesel generator exhaust. The signals with the mass range $m/z = 75$–300 are measured and analyzed. The sharp rise of benzene and naphthalene is shown at the engine start. It was also demonstrated that good agreement was observed between results obtained with REMPI-TOFMS and conventional extractive sampling. A similar method has been applied to a real-time measurement of trace aromatics during operation of aircraft ground equipment.

LI-TOFMS has the potential to analyze particle constituents. Figure 7.10 shows an example of a direct inlet aerosol analysis using laser desorption/ionization TOFMS.[7.11] The particle laser desorption is carried out by a CO_2 laser at 10.6 μm,

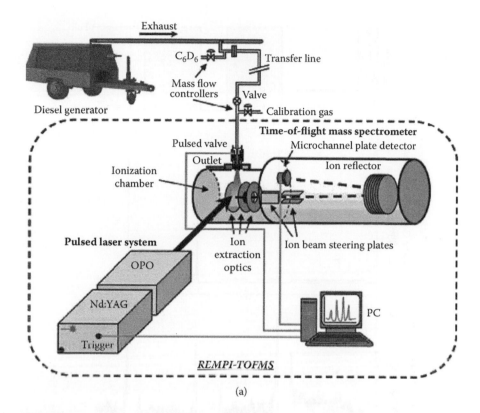

(a)

FIGURE 7.9

Application of REMPI-TOFMS to diesel generator exhaust. (a) Schematic of the experimental setup. The major components of the REMPI-TOFMS instrument are a Nd:YAG + OPO laser with UV frequency doubling and a time-of-flight mass spectrometer. The major sampling components as used in the diesel generator exhaust sampling are indicated. (b) Transient REMPI-TOFMS results. (Top frame) Loading profile of diesel generator during warm restart test schedule. Three conditions are encountered for the diesel generator: full load, zero load, and off position. (Middle frame) Transient benzene emissions following three restarts of the diesel generator as recorded at the wavelength for benzene detection. (Bottom frame) Transient naphthalene emissions recorded simultaneously. A 25% increase in emission level is also observed between zero and full load of diesel generator. (Reprinted with permission from L. Oudejans, A. Touati, and B.K. Gullett, "Real-time, on-line characterization of diesel generator air toxic emissions by resonance-enhanced multiphoton ionization time-of-flight mass spectrometry," *Analytical Chemistry*, 76(9), 2517–2524, 2004. Copyright 2004 American Chemical Society.)

and a KrF excimer laser at 248 nm is used for the ionization. Particles produced by wood combustion and car engines have been measured using laser desorption/ionization systems. The mass patterns of diesel engine and wood stove exhausts are shown in Figure 7.10(b) and 7.10(c), respectively. There were significant differences between the two patterns, and it was possible to distinguish the particles from different sources by their mass patterns.

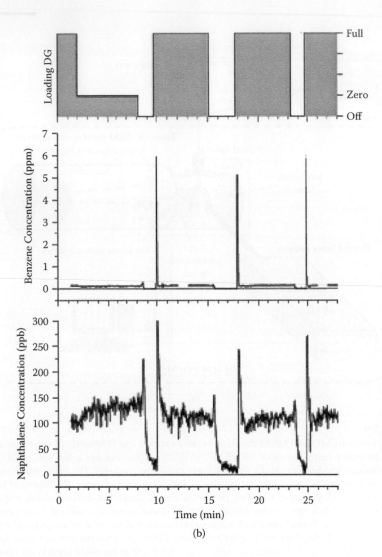

FIGURE 7.9
(Continued)

LI-TOFMS has also been utilized to detect constituents classified as super-fine particles—so-called nanoparticles—from diesel engines.[7.12] The experimental apparatus for classified nanoparticle constituent detection is shown in Figure 7.11. The sampled gas was introduced into the dynamic mechanical analysis (DMA) for the classification of nanoparticles and corresponding number density measurement using a condensation particle counter. The classified particles were transferred and heated inside a stainless steel pipe, and the vaporized gaseous components of the heated nanoparticles were introduced into the vacuum chamber of the LI-TOFMS system. The gas in the

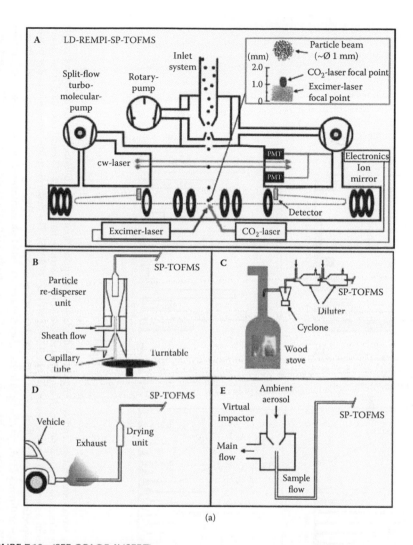

(a)

FIGURE 7.10a (SEE COLOR INSERT)

Analyses of aerosol particles produced by wood combustion and car engines using laser desorption/ionization TOFMS. Schematic of the experimental setup. (A) Sketch of the bipolar SP-TOFMS instrument. Top right, the focal point of the CO_2 laser (desorption laser, λ 10.6 µm) and ionization laser (ionization laser, λ 248 nm) visualized with thermal paper that was situated in the middle of the SP-TOFMS ionization region. The size of the particle beam is indicated (particle beam profile is shown as the top view, whereas laser profiles were measured in a direction orthogonal to the particle beam. (B) Sketch of a small-scale powder disperser. (C) Sketch of a wood stove with a cyclone and diluter. (D) Sketch of car exhaust measurement with a denuder drying unit. (E) Principal sketch of a virtual impact or aerosol concentrator sampling setup used for ambient measurement. (Reprinted with permission from M. Bente, M. Sklorz, T. Streibel, and R. Zimmermann, "Online laser desorption-multiphoton postionization mass spectrometry of individual aerosol particles: Molecular source indicators for particles emitted from different traffic-related and wood combustion sources," *Analytical Chemistry*, 80(23), 8991–9004, 2008. Copyright 2008 American Chemical Society.)

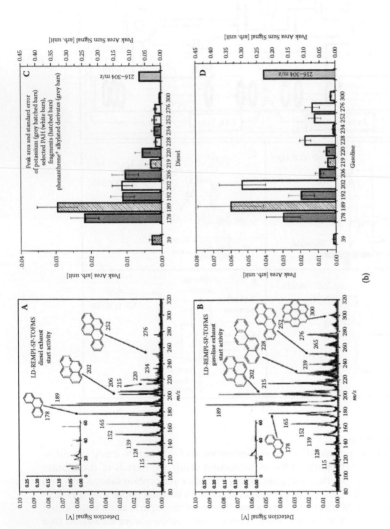

(b)

FIGURE 7.10b

Analyses of aerosol particles produced by wood combustion and car engines using laser desorption/ionization TOFMS. Averaged single-particle mass spectra of car exhaust aerosol from a diesel and a gasoline car. (A and B) Averaged single-particle mass spectra (15 each) obtained from an on-line measurement of car exhaust aerosol from a diesel (A) and a gasoline (B) car (LD-REMPI-SP-TOFMS). The desorption and ionization laser power densities are 5×10^7 and 5×10^6 W/cm², respectively. (B and C) Peak areas with standard errors of potassium (light gray hatched bar), phenanthrene and larger PAHs (white bars), alkylated phenanthrenes (dark gray bars), selected fragments of alkylated phenanthrenes (white hatched bars) as well as the sum of the large mass peak area (216–304 m/z, light gray bar) from the diesel (C) and gasoline (D) car exhaust experiment. (Reprinted with permission from M. Bente, M. Sklorz, T. Streibel, and R. Zimmermann, "Online laser desorption-multiphoton postionization mass spectrometry of individual aerosol particles: Molecular source indicators for particles emitted from different traffic-related and wood combustion sources," *Analytical Chemistry, 80(23), 8991–9004, 2008. Copyright 2008 American Chemical Society.)*

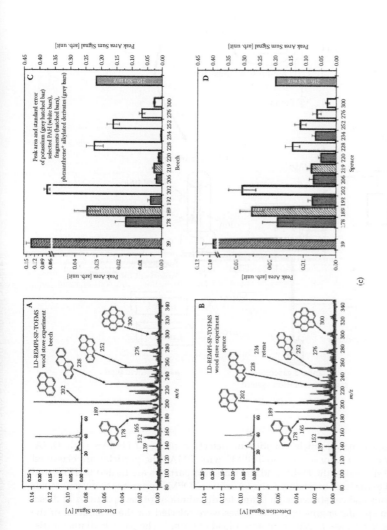

(c)

FIGURE 7.10c

Analyses of aerosol particles produced by wood combustion and car engines using laser ionization TOFMS. Averaged single-particle mass spectra of the combustion aerosol from beech and spruce wood burning. (A and B) Averaged single-particle mass spectra (50 each) obtained from an on-line measurement of the combustion aerosol from beech (A) and spruce (E) wood burning (LD-REMPI-SP-TOFMS). The desorption and ionization laser power densities are 5×10^7 and 5×10^6 W/cm² , respectively. (C and D) Peak areas with standard errors of potassium (light gray hatched bar), phenanthrene and larger PAHs (white bars), alkylated phenanthrenes (dark gray bars), selected fragments of alkylated phenanthrenes (white hatched bars) as well as the sum of the large mass peak area (216–304 m/z, light gray bar) from the beech (C) and spruce (D) wood burning experiment. (Reprinted with permission from M. Bente, M. Sklorz, T. Streibel, and R. Zimmermann, "Online laser desorption-multiphoton postionization mass spectrometry of individual aerosol particles: Molecular source indicators for particles emitted from different traffic-related and wood combustion sources," *Analytical Chemistry*, 80(23), 8991–9004, 2008. Copyright 2008 American Chemical Society.)

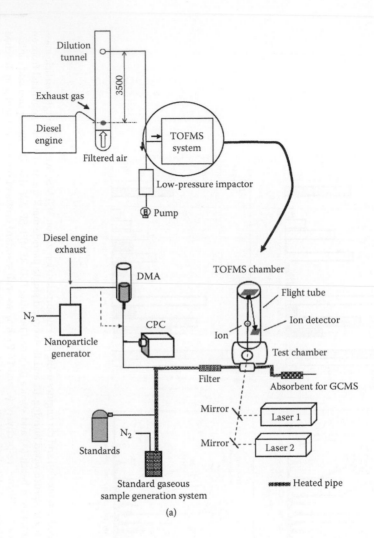

(a)

FIGURE 7.11

Apparatus for classified nanoparticle constituent detection using LI-TFMAS with DMA. Experimental apparatus for laser ionization TOFMS with DMA. The particles are classified to a specific particle size using the DMA, and the classified particles are heated to 473 K and directly transported to the laser breakdown and ionization TOFMS apparatus to measure their desorbed compositions and constituents. (a) Diameter distributions of nanoparticles in diesel engine exhaust. Two specific engine operation cases were selected for the evaluation of the TOFMS system. One is the case at 2000 rpm and a load of 0 nm^{-1}. The particle diameter distribution under these conditions peaked at 20–40 nm, which corresponds to the nuclei mode. The other is the case at 1000 rpm and a load of 30 nm^{-1}. Here, the particle diameter distribution peaked at 80–100 nm, showing an accumulation mode. (b) Identification of PAHs in the mass spectrum of classified 30-nm nanoparticles. Analysis of the mass spectra shows that the nanoparticles contained not only the standard PAHs such as anthracene and benzo(a)pyrene, but also several types of PAHs with methyl, ethyl, and so on. (Reprinted from Y. Deguchi, N. Tanaka, M. Tsuzaki, A. Fushimi, S. Kobayashi, and K. Tanabe, _Environmental Chemistry_, 5(6), 402, 2008. With permission.)

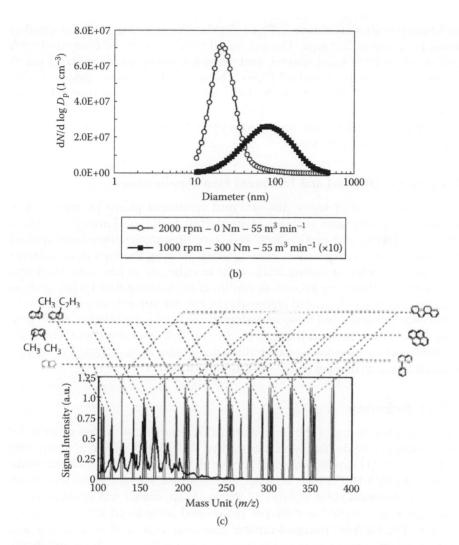

(b)

(c)

FIGURE 7.11
(Continued)

chamber was ionized by unfocused laser irradiation with 15 mJ per pulse at 266 nm for REMPI and by focused laser irradiation with 500 mJ per pulse at 1064 nm using a 150-mm lens for breakdown. The ions generated were transferred to a reflectron-type TOFMS chamber and detected by a MCP detector. A 4-cycle, 8-liter direct-injection diesel engine was used for the nanoparticle measurement. Engine exhaust was diluted using a dilution tunnel with a polished surface. The nanoparticles were also sampled by a low-pressure impactor and analyzed by means of thermal desorption GC-MS for comparison with the TOFMS results. The results of classified particle measurements of 30-nm particles are shown in Figure 7.11(b). The signal appeared at a higher mass range up to 300 amu. The measured mass spectra are compared with the standard PAH mass spectra, and several mass spectra are identified as corresponding to the standard PAHs. These results are also compared with the measurement results from the low-pressure impact and thermal conditions using the equilibrium of bipolar charge distribution. The two results show the same order concentrations for these PAHs, which demonstrates the validity of the TOFMS system.

7.3.2 Waste Disposal and Treatment Plant Applications

Emissions control of waste disposal and treatment plants is important to ensure the safety and security of industrial facilities. Emissions include NO_x, CO, PAHs, and particulates, and laser diagnostics have been applied to monitor these emissions in various ways. Among these emissions, larger organic molecules including PAHs exist in exhausts at low concentrations, and some of them are known as carcinogens. Chlorinated PAHs, such as dioxins and polychlorinated biphenyls (PCBs), are super-toxic even at a very low concentrations, and these emissions have to be controlled in a safe way. LI-TOFMS is the method that can be accessible to monitor these molecules on-line, and its monitoring capabilities have been demonstrated in several industrial waste disposal[7.7],[7.13] and treatment plants.[7.5]

7.3.2.1 Incinerator

An incinerator is one of the facilities most frequently used in waste disposal and treatment fields. Emission monitoring and control of PAHs along with chlorinated PAHs are important to cope with the reduction of toxic materials. On-line REMPI-TOFMS has been applied to an industrial 22-MW hazardous waste incineration plant. Figure 7.12 shows its apparatus and applications.[7.7] Flue gas was sampled at different places and introduced into the TOFMS system. The Nd:YAG pumped tunable dye laser was used for selective and sensitive detection of mono- and polycyclic aromatic hydrocarbons and mono-chlorobenzene (MCB). Several laser wavelengths were used for the sensitive detection of each molecule, such as 269.82 nm for MCB. The achieved detection limit reached in the ppt (parts per trillion) concentration range.

7.3.2.2 PCB Disposal Plant

PCBs are a type of chlorinated aromatic hydrocarbons that were widely used in condensers and other applications prior to 1970. They are now considered hazardous environmental pollutants. PCB processing requires treatment of PCBs, as well as cleaning of PCB containers and processing of contaminated paper, wood, and other exposed or contaminated substances. It must be ensured in the PCB treatment process that PCBs are not released into

(a)

FIGURE 7.12a
Application of REMPI-TOFMS to industrial waste incineration plant. Schematic of the experimental setup. (a) Photograph of the mobile REMPI-TOFMS laser mass spectrometer. The main parts of the system are (1) Nd:YAG pumped dye laser; (2) reflectron-TOF mass spectrometer; (3) sample inlet system; (4) transfer line; (5) ionization chamber; (6) digital storage oscilloscope; (7) electronics and vacuum control units; (8) Nd:YAG power supply; (9) calibration gas generator for external quantification; (10) power supply for heating devices. (b) Photograph of the mobile REMPI-TOFMS laser mass spectrometer at the measurement place at the 22-MW hazardous waste incinerator. The two measurement places at the beginning and the end of the boiler section are indicated by the numbers 2 and 3, respectively. (Reprinted with kind permission from Springer Science+Business Media: *Fresenius' Journal of Analytical Chemistry*, "Direct observation of the formation of aromatic pollutants in waste incineration flue gases by on-line REMPI-TOFMS laser mass spectrometry," 366(4), 2000, 368–374, R. Zimmermann, H.J. Heger, A. Kettrup, and ·U. Nikola.)

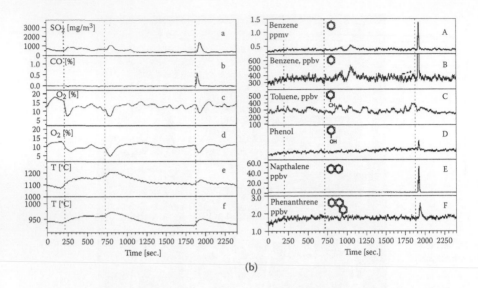

(b)

FIGURE 7.12b

Application of REMPI-TOFMS to industrial waste incineration plant. Online measurement of chemical species in the flue gas and other parameters by (left) conventional sensors and of organic trace compounds by (right) REMPI-TOFMS laser mass spectrometry. During non-stationary combustion condition caused by barrel uptake processes: (left) conventional sensors: (a) sulfur dioxide, (b) carbon monoxide, (c) oxygen in the rotary kiln, (d) oxygen in the afterburning chamber, (e) temperature in the rotary kiln, and (f) temperature in the after-burning chamber; (right) REMPI-TOFMS measurement: (A) benzene, (B) benzene enlarged, (C) toluene, (D) phenol, (E) naphthalene, and (F) phenanthrene. (Reprinted with kind permission from Springer Science+Business Media: *Fresenius' Journal of Analytical Chemistry*, "Direct observation of the formation of aromatic pollutants in waste incineration flue gases by on-line REMPI-TOFMS laser mass spectrometry," 366(4), 2000, 368–374, R. Zimmermann, H.J. Heger, A. Kettrup, and ·U. Nikola.)

exhaust gas or the work environment atmosphere. The conventional analysis method requires 2~3 days for gas sampling, and the concentration process and faster PCB monitoring methods are needed for safe management.

The apparatus of the PCB disposal plant application is shown in Figure 7.13.[7.5] The Nd:YAG laser, having a wavelength of 266 nm and a pulse width of 100 ps, is used for PCB ionization to reduce PCB fragmentation. Ionized molecules were introduced into the ion trap cell using a dynamic trapping method. Figure 7.13(b) shows measurement results using nanosecond and picosecond laser ionizations for PCBs introduced into the apparatus. Ionization by laser light with a pulse width of 100 ps clearly results in a stronger PCB signal than that of laser light with a pulse width of 5 ns. By using the picosecond pulse width, the PCB signal increases over 10 times for (2~4 Cl) PCBs. These results also suggest that ionization using laser light with ps pulse width reduces the fragmentation of PCBs. Figure 7.13(c) presents the container treatment process in the disposal plant for PCBs, and Figure 7.13(d) shows an example of results measured in the work environment atmosphere of the PCB disposal plant

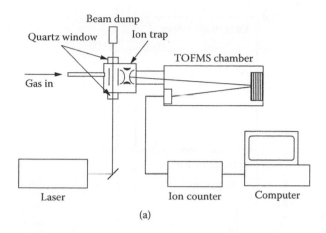

(a)

FIGURE 7.13

Application of LI-TOFMS to PCB disposal plant. (a) Schematic diagram of the PCB monitoring apparatus. The apparatus was composed of a picosecond laser, an ion trap TOFMS device, an ion counter, and a computer. The operation of this system was automated for plant monitoring use. (b) PCB measurement results using nanosecond and picosecond laser ionization. The laser pulse energies were set to the same level (10 mJ/pulse) in both cases. The use of a shorter-pulse laser ionization source reduces the fragmentation of PCBs and enhances the detection sensitivity. (c) The container treatment process in the disposal plant for PCBs. (d) Example of the analysis results of the work environment atmosphere of the PCB disposal plant obtained by LI-IT-TOFMS. The concentration of PCBs in the atmosphere of the work environment of the plant, which was substantially lower than the 0.10 mg/m³ N specified by statutory regulations, could be analyzed within 1 min. It was possible to make continuous long-term measurements for over 2000 h (120,000 data points). [(a) and (b) Reprinted from Y. Deguchi , S. Dobashi, N. Fukuda, K. Shinoda, and M. Morita, "Real time PCB monitoring using time of flight mass spectrometry with picosecond laser ionization," *Environmental Science and Technology*, 37(20), 4737–4742. Copyright 2003 American Chemical Society. (c) and (d) Reprinted with kind permission from Springer Science+Business Media, *Journal of Material Cycles and Waste Management*, "Safety management by use of laser mass spectrometry of polychlorinated biphenyls (PCBs) in the processed gas and work environment of a PCB disposal plant," 11(2), 2009, 148–154, S. Dobashi, Y. Yamaguchi, Y. Izawa, A. Wada, and M. Hara.]

obtained by LI-IT-TOFMS.[7.13] As indicated in Figure 7.13, the PCB concentration was maintained under 0.1 mg/nm³, which has been specified by statutory regulations. It was demonstrated that it is possible to perform on-line analysis in the working environment for over 2,000 hours.

7.5 Future Developments

There are other LI-TOFMS applications in several fields. TOFMS has been mainly developed in medical and pharmaceutical fields, and main applications are still in these fields. Although LI-TOFMS has great features such

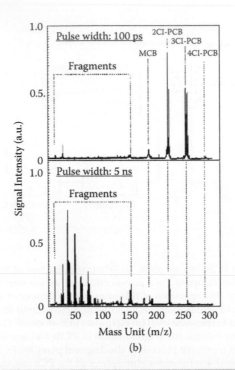

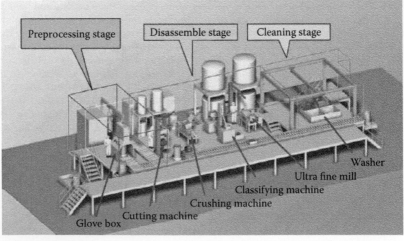

(c)

FIGURE 7.13
(Continued)

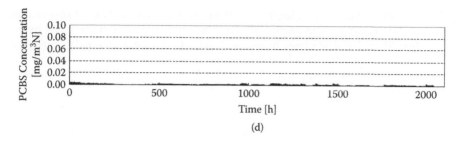

FIGURE 7.13
(Continued)

as a fast response and super-high sensitivity, its high cost limits its application to a specific area. LI-TOFMS requires a TOFMS system with vacuum pumps and high-power lasers, and they are the main contributors to cost and size. Since LI-TOFMS has the potential to monitor industrial facilities—for example, it was demonstrated as a PCB monitoring system for more than 2000 h[7.13]—advancement of lasers and TOFMS devices in size and cost will enhance its applicability. There are also several advances in this technology, which include a single photon ionization method and combination of other methods such as gas chromatography and laser desorption. These advancements will pave the way to new application fields of LI-TOFMS.

References

[7.1] D.M. Lubman, *Lasers and mass spectrometry*, New York, Oxford University Press, 1990.

[7.2] D.S. Zakheim and P.M. Johnson, "Rate equation modeling of molecular multiphoton ionization dynamics," *Chemical Physics*, 46(3), 263–272, 1980.

[7.3] R. Zimmermann, U. Boesl, C. Weickhardt, D. Lenoir, K.-W. Schramm, A. Kettrup and E.W. Schla, "Isomer-selective ionization of chlorinated aromatics with lasers for analytical time-of-flight mass spectrometry: First results of polychlorinated dibenzo-p-dioxins (PCDD), biphenyls (PCB) and benzenes (PCBz)," *Chemosphere*, 29(9), 1877–1888, 1994.

[7.4] H. Oser, R. Thanner, and H-H. Grotheer, "Continuous monitoring of ultratrace products of incomplete combustion during incineration with a novel mobile JET-REMPI device," *Chemosphere*, 37(9), 2361–2374, 1998.

[7.5] Y. Deguchi , S. Dobashi, N. Fukuda, K. Shinoda, and M. Morita, "Real time PCB monitoring using time of flight mass spectrometry with picosecond laser ionization," *Environmental Science and Technology*, 37(20), 4737–4742, 2003.

[7.6] S. Dobashi, Y. Yamaguchi, Y. Izawa, Y. Deguchi, A. Wada and M. Hara, "Laser mass spectrometry: Rapid analysis of polychlorinated biphenyls in exhaust gas of disposal plants," *Journal of Environment and Engineering*, 2(1), 25–34, 2007.

[7.7]   R. Zimmermann, H.J. Heger, A. Kettrup, and ·U. Nikola, "Direct observation of the formation of aromatic pollutants in waste incineration flue gases by on-line REMPI-TOFMS laser mass spectrometry," *Fresenius' Journal of Analytical Chemistry*, 366(4), 368–374, 2000.

[7.8]   L. Oudejans, A. Touati, and B.K. Gullett, "Real-time, on-line characterization of diesel generator air toxic emissions by resonance-enhanced multiphoton ionization time-of-flight mass spectrometry," *Analytical Chemistry*, 76(9), 2517–2524, 2004.

[7.9]   B.K. Gullett, A. Touati, L. Oudejans, and S.P. Ryan, "Real-time emission characterization of organic air toxic pollutants during steady state and transient operation of a medium duty diesel engine," *Atmospheric Environment*, 40(22), 4037–4047, 2006.

[7.10]   B. Gullett, A. Touati, and L. Oudejans, "Use of REMPI–TOFMS for real-time measurement of trace aromatics during operation of aircraft ground equipment," *Atmospheric Environment*, 42(9), 2117–2128, 2008.

[7.11]   M. Bente, M. Sklorz, T. Streibel, and R. Zimmermann, "Online laser desorption-multiphoton postionization mass spectrometry of individual aerosol particles: Molecular source indicators for particles emitted from different traffic-related and wood combustion sources," *Analytical Chemistry*, 80(23), 8991–9004, 2008.

[7.12]   Y. Deguchi, N. Tanaka, M. Tsuzaki, A. Fushimi, S. Kobayashi, and K. Tanabe, "Detection of components in nanoparticles by resonant ionization and laser breakdown time-of flight mass spectroscopy," *Environmental Chemistry*, 5(6), 402–412, 2008.

[7.13]   S. Dobashi, Y. Yamaguchi, Y. Izawa, A. Wada, and M. Hara, "Safety management by use of laser mass spectrometry of polychlorinated biphenyls (PCBs) in the processed gas and work environment of a PCB disposal plant," *Journal of Material Cycles and Waste Management*, 11(2), 148–154, 2009.

8

Advances in Laser Diagnostics and Their Medical Applications

8.1 Advances in Laser Diagnostics

There are numerous measurement methods using laser diagnostics, which include measurement technologies of temperature, concentration, velocity, density, pressure, distance, shape, particle size, and so on. These methods utilize the characteristics of lasers: coherent (directional), monochromatic, and short-pulse features. Many of these have been commercialized and utilized in various industrial fields. These measurement technologies rely heavily on the advancement of lasers and associated products, including detectors. Advances in femtosecond (fs) lasers have introduced new fields of measurement technologies such as THz technology (see Section 8.3). Optical fiber sensors, in which optical fibers themselves are sensors, have been developed to detect temperature, strain, vibration, and so on. A Brillouin optical-time-domain reflectometer (BOTDR) is an example of optical fiber sensors and it can measure strain along arbitrary regions of an optical fiber. This method has been used to measure strain on large structures. There have also been advances in laser diagnostics to achieve higher-dimensional measurements. Laser Doppler velocimetry (LDV) has been widely used to measure velocity. In LDV, two laser beams are focused to cross with each other at their focal point, where they interfere and generate a set of straight fringes. When a particle passes through these fringes and the scattered light is measured by a detector, the signal has a beat with frequency related to the velocity of the particle. The velocity can be calculated from the beat frequency. LDV is a point measurement technology, and particle image velocimetry (PIV) has developed as a two-dimensional (2-D) velocity measurement method. PIV measures the motion of particles using a double-pulsed laser and the motion of particles during double laser pulses is used to calculate 2-D velocity information of the flow.

In this chapter, rapidly advancing laser diagnostics are briefly discussed in the following sections: optical coherence tomography (OCT) in Section 8.2 and THz technology in Section 8.3. Up to this point, laser-induced fluorescence (LIF), laser-induced breakdown spectroscopy (LIBS), spontaneous Raman spectroscopy, coherent anti-Stokes Raman spectroscopy (CARS), tunable diode laser absorption spectroscopy (TDLAS), and laser ionization time-of-flight mass spectrometry (LI-TOFMS) have been mainly discussed from the point of view of industrial applications. There has also been a great deal of progress in medical applications of these technologies, which are briefly summarized in Section 8.4.

8.2 Optical Coherence Tomography

Optical coherence tomography (OCT) is an interferometric technology that detects backscattered light from an object using low-interference laser light. OCT provides information of the backscattered light intensity as a function of depth from the object surface. The basic principle of OCT is shown in Figure 8.1. Light from a low-coherence broad-bandwidth light source is split into two beams that are used as sample and reference light. Sample light entering the object is backscattered with different time delays according to the object's structure, that is, the depth from the object surface. The backscattered light and reference light are recombined to form an interference pattern by moving the mirror of reference light. The interference pattern only arises when the path difference of signal and reference light lies within the coherence length of light. Though OCT is a 1-D measurement technology, it is easy to construct a 3-D image of the object structure by scanning the light two-dimensionally. The axial resolution of OCT is inversely proportional to the spectral bandwidth of a light source and

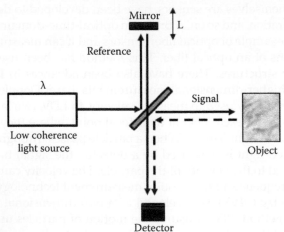

FIGURE 8.1
Basic principle of OCT. OCT is an interferometric technology which detects a backscattered light from an object using low interference laser light, and it provides information of the backscattered light intensity as a function of depth from the object surface. The axial resolution often reaches to 1-2 μm, while the lateral resolution of OCT depends on the optics used in a measurement system and it becomes around 20 μm.

is equivalent to the coherence length of the light source ℓ_c. The axial resolution Δz is given by the following equation:

$$z = \ell_c$$

$$= \frac{2 \ln 2}{\pi} \frac{\lambda_0^2}{\lambda} \tag{8.1}$$

where $\Delta\lambda$ is the spectral bandwidth and λ_0 is the central wavelength of light. The axial resolution often reaches 1–2 μm. The lateral resolution of OCT depends on the optics used in the measurement system, and OCT usually possesses a lateral resolution around 20 μm. Therefore the lateral resolution often limits the measurement resolution of OCT. The imaging depth of OCT is limited to 1–2 mm below the surface of the object, and OCT has been used to measure a shallow structure from the object's surface.

Several types of OCTs have been developed and some of these are shown in Figure 8.2. They are often called time domain OCT (TD-OCT), frequency

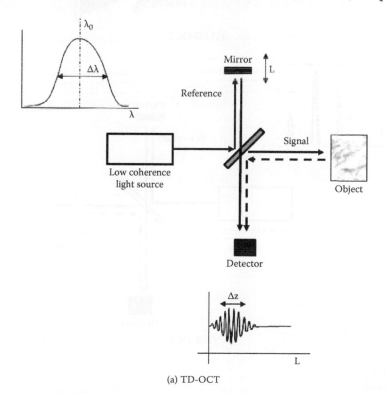

(a) TD-OCT

FIGURE 8.2 (SEE COLOR INSERT)
Several types of OCTs. There are mainly three types of OCTs. In TD-OCT the interference pattern arises according to time delays of scattered light, while the interference pattern is detected as a spectrum in FD-OCT. SS-OCT uses a swept laser sources at a high frequency scanning rate. The advantage of SS-OCT is a high signal-to-noise signal. (a) TD-OCT, (b) FD-OCT, (c) SS-OCT.

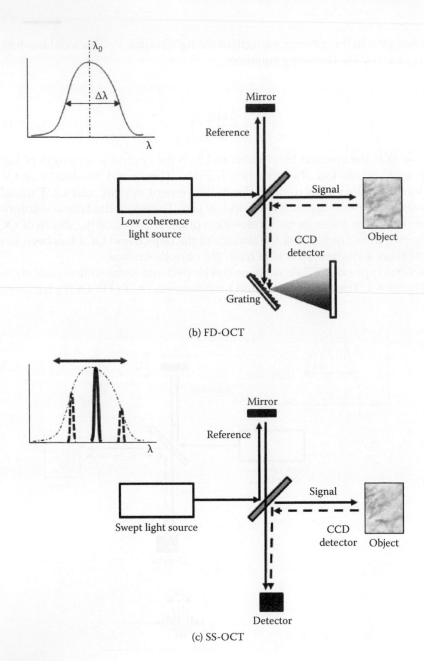

(b) FD-OCT

(c) SS-OCT

FIGURE 8.2
(Continued)

domain OCT (FD-OCT), and swept source OCT (SS-OCT). TD-OCT is a method shown in Figure 8.2(a) and the interference pattern arises according to time delays of scattered light. In FD-OCT the interference pattern is detected as a spectrum using a grating and a charge-coupled device (CCD) detector, as shown in Figure 8.2(b). In this method it is not necessary to move the mirror of reference light. SS-OCT uses a swept laser source at a high-frequency scanning rate (typically few dozens of kHz), as shown in Figure 8.2(c). The advantage of this method is a high signal-to-noise signal.

There have been several commercially available OCT systems and they have been employed mainly in medical applications for eyes,[8.1] teeth,[8.2],[8.3] skin,[8.4] heart,[8.5] and organs and tissues.[8.1],[8.6] Especially in ophthalmology, OCT is a valuable tool to obtain detailed retina images. Figure 8.3 shows monitoring results of early caries development using Raman spectroscopy and optical coherence tomography.[8.2] The axial and transverse resolutions of OCT were 10–20 μm. Both OCT and polarized Raman spectroscopy have the potential to differentiate healthy from early diseased enamel. Combinations of these measurements can offer reasonable precision with good repeatability and reproducibility.

There are also some industrial applications of OCT. OCT can offer good axial and transverse resolutions, and the method can be applied to industrial

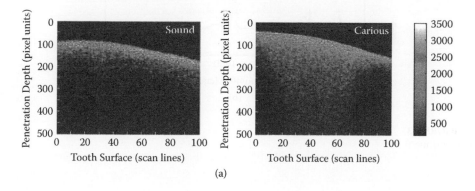

(a)

FIGURE 8.3

Monitoring of early caries development using Raman spectroscopy and optical coherence tomography. (a) OCT measurement results. 2D OCT images from the same tooth acquired from a region of sound enamel (left) and carious enamel (right). (b) Raman spectra measurement results. Representative parallel-polarized and cross-polarized Raman spectra obtained from a sound enamel and b carious enamel. The prominent phosphate v1 PO_4^{3-} vibration at approximately 959 cm^{-1} is highlighted. (Reprinted with kind permission from **Springer Science+Business Media**: *Analytical and Bioanalytical Chemistry*, "Precision of Raman depolarization and optical attenuation measurements of sound tooth enamel," 387(5), 1613–1619, M.G. Sowa et al., copyright 2007.)

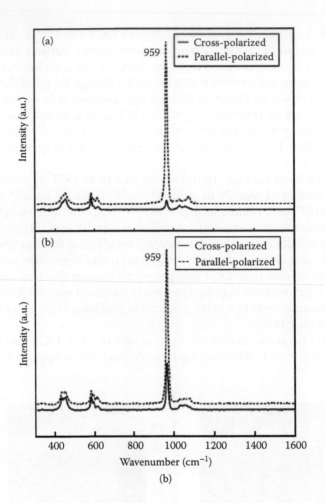

FIGURE 8.3
(Continued)

inspections[8.7],[8.8] and process controls.[8.9] Figure 8.4 shows OCT measurements of a circuit feature.[8.7] The OCT system employed a femtosecond (fs) supercontinuum laser source and a balanced detector in a time-domain OCT configuration, achieving an axial resolution 700 nm in silicon. A sequence of metallization interconnections located at the perimeter of the device was observed on the OCT image as a periodic array of high- and low-reflectivity regions.

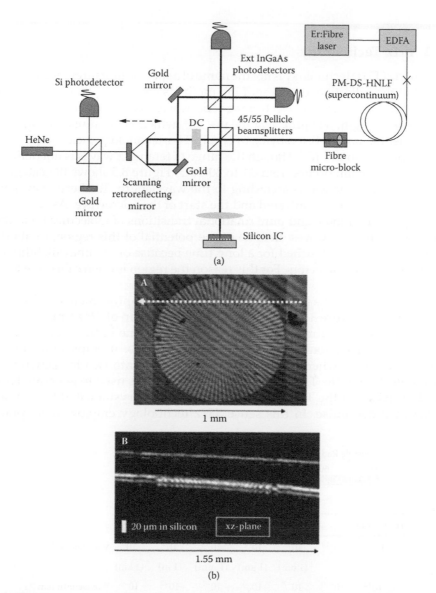

FIGURE 8.4
Sub-surface inspection of silicon integrated circuits using optical coherence tomography. (a) Experimental configuration utilizing a supercontinuum source, balanced-detection and a HeNe-laser-calibrated reference arm. (b) LSM image of a circuit feature on a device-under-test: (A) LSM image of a circuit feature on a device-under-test exposing radial fabrication distribution and (B) a magnified OCT image of this feature confirming this distribution. The OCT image was acquired along the scan direction indicated by the red dotted line. (For interpretation of the references to color in this figure legend, the reader is referred to the web version of this article.) (Reprinted from *Microelectronic Engineering*, 87(9), K.A. Serrels, M.K. Renner, and D.T. Reid, "Optical coherence tomography for non-destructive investigation of silicon integrated-circuits," 1785–1791, Copyright 2010, with permission from Elsevier.)

8.3 THz Technology

Terahertz (THz) technology is not the name of a specific measurement technology but the collective term describing methods employing terahertz waves. Terahertz waves are the electromagnetic waves with frequency around 1 THz (wavelength: 300 μm). *Tera* means 10^{12}, which is often used as unit of computer memory (TB, terabyte); "THz" in THz technology means light with frequency of around 10^{12}. Though the range of terahertz waves is not defined clearly, its domain occurs from 0.1 to 30 THz. Figure 8.5 shows the category of electromagnetic waves according to the wavelength. Terahertz waves lie between the end of the infrared and the start of the microwave. As shown in the figure, vibrational and pure rotational transitions of molecules lie in the region of terahertz waves. In spite of the potential of this region, terahertz waves remained untouched for a long time because of the unavailability of light sources and detectors. For this reason the region is often called the terahertz gap.

THz technology has greatly advanced due to improvements in fs-laser technologies. Figure 8.6 shows a typical optical setup of THz time-domain spectroscopy. THz time-domain spectroscopy is one of the types of generation and detection methods of terahertz waves. It employs a pump-probe technique where the fs-laser beam is split into two for generation and detection of the THz signal. The pump beam is used to generate terahertz waves and the THz signal is detected by a crystal gated by a time-delayed probe pulse. In this sense THz technology employs absorption

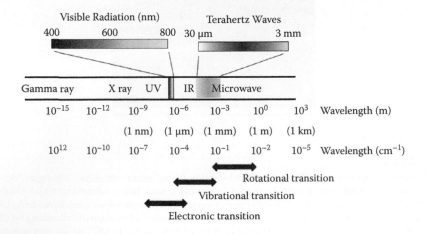

FIGURE 8.5 (SEE COLOR INSERT)
Category of electromagnetic waves according to the wavelength. Terahertz waves lie between the end of the infrared and the start of the microwave. Vibrational and pure rotational transitions of molecules lie in this region. In spite of the potential of this region, terahertz waves remained untouched for a long time, and the region is often called terahertz gap.

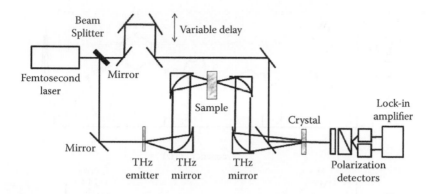

FIGURE 8.6
Typical optical setup of THz time-domain spectroscopy. THz time-domain spectroscopy is one of the types of generation and detection methods of terahertz waves. It employs a pump-probe technique where the fs-laser beam is split into two for generation and detection of the THz signal. The pump beam is used to generate terahertz waves and the THz signal is detected by a crystal gated by a time-delayed probe pulse.

spectroscopy to evaluate a transmitted signal. Because vibrational and pure rotational transitions of molecules lie in the region of terahertz waves, this method has been applied to detect gases such as NH_3[8.10] and H_2O.[8.11] However, the method is not mature in terms of industrial applications compared to TDLAS. It is important to know that atmospheric attenuation of terahertz waves arises mainly due to water vapor and this prevents long-distance applications using terahertz waves compared to other laser diagnostics, including TDLAS.[8.12]

One of the attractive characteristics of terahertz waves for advanced applications is that terahertz waves can penetrate into many nonmetal materials. Using this feature of terahertz waves, it is possible to see through objects. Paper, cardboard, plastic, and clothing materials are transparent in the range of terahertz waves, and there have been extensive studies in security applications, especially in explosive detection fields. THz absorption spectra of explosives RDX, HMX, PETN, and TNT are shown in Figure 8.7.[8.13] It is obvious that each explosive has its unique spectrum, and it is distinguishable from the others. The great merit of THz technology is its ability to detect these explosives covered by papers or plastics. In this sense THz technology has an advantage over LIBS, which has to measure these samples directly (see Section 4.3.3).

There has also been a good deal of research on THz spectroscopy in security applications,[8.12],[8.13] pharmaceutical applications,[8.12],[8.14] and medical applications.[8.12],[8.15],[8.16] THz technology can also be applied to detect three-dimensional (3-D) information of materials using a tomographic analysis. Figure 8.8 shows an example of 3-D image of soot in a ceramic filter model. [8.17] It is desirable to monitor soot that accumulates inside a filter in a non-destructive manner. A 3-D image of soot can be successfully reconstructed

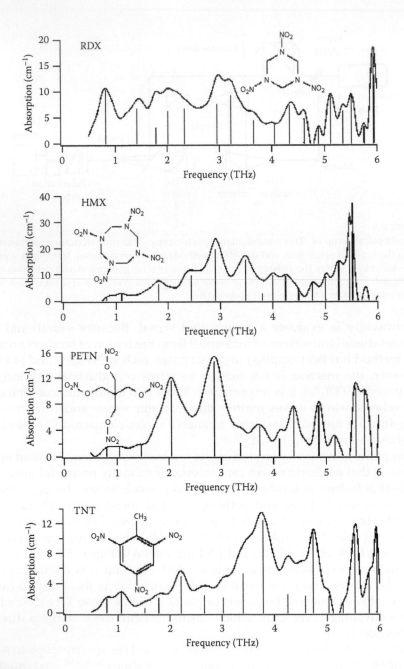

FIGURE 8.7
THz absorption spectra of explosives; RDX, HMX, PETN, and TNT. The chemical structures of the explosives are also shown. Lorentzian lineshapes were fitted to the spectra and indicated by lines at the center frequency. (Reprinted from *Chemical Physics Letters*, 434(4–6), M.R. Leahy-Hoppa et al., "Wideband terahertz spectroscopy of explosives," 227–230, Copyright 2007, with permission from Elsevier.)

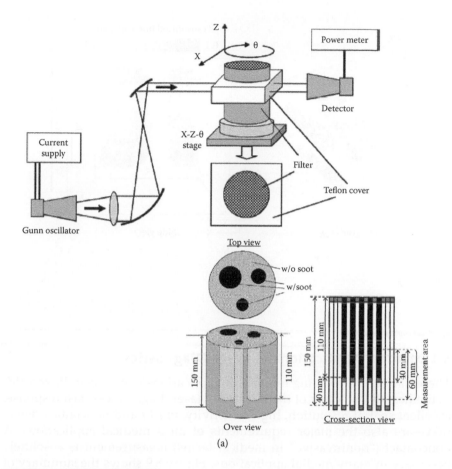

FIGURE 8.8 (SEE COLOR INSERT)
3-D image of soot in a ceramic filter model. (a) Schematic diagram of the 3D-CT image acquisition system and the test sample of the soot removal filter. Soot accumulates in the three cylindrical areas with different radii. (b) The 3D-CT reconstruction image. The regions of accumulated soot can be clearly seen. The soot distribution in the X-Y plane can be confirmed, though the resolution of the image is little depleted. Some loss of resolution and distortion of the image occurs because of diffraction due to the mismatch of the refractive index of the filter and the Teflon cover, and due to artifacts of the inverse-Fourier transform. (Reprinted from *Comptes Rendus Physique*, 11(7–8), K. Kawase et al., "THz imaging techniques for nondestructive inspections," 510–518, Copyright 2010, with permission from Elsevier.)

through a ceramic filter using terahertz waves. Although THz technology is a new and attractive measurement field, its application is still limited compared to other laser-based measurement technologies. The main reason is that measurement devices often consist of fs-lasers and crystal-based detectors. There has been great advancement in fs-laser technologies, but fs-lasers are still an expensive and vulnerable tool for industrial applications. Reasonable, compact, and powerful THz sources and detectors[8.18] will lead the technology to industrial fields.

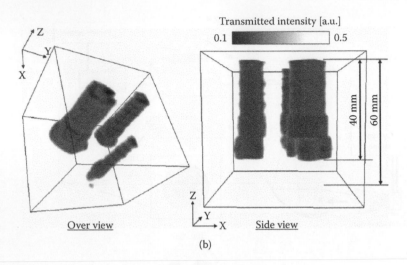

FIGURE 8.8
(Continued)

8.4 Medical Applications of Laser Diagnostics

There has been a growing number of applications using laser diagnostics in medical fields. Many of the features of laser diagnostics—fast response, excellent spatial resolution, high sensitivity, *in situ* and noncontact detection—are also the major requirements of most medical applications. A noncontact ("noninvasive" in medical terms) measurement is absolutely necessary in many medial applications. Figure 8.9 shows the summary of medical applications using LIF, LIBS, spontaneous Raman spectroscopy, CARS, TDLAS and LI-TOFMS. These applications extend from *in vivo* measurements of live cells to biological material analyses. In medical applications, *in vivo* is often used to refer to experimentation using a living organism; *in vivo* is also used for experimentation done in live isolated cells. It can be said that the term *in vivo* corresponds to *in situ* in industrial applications. OCT and THz technologies have been also applied to medical fields, and these applications are briefly described in Sections 8.2 and 8.3, respectively.

8.4.1 Laser-Induced Fluorescence

As discussed in Chapter 3, LIF has been applied to two-dimensional (2-D) detection of temperature and species concentration due to its strong signal intensity. There are two types of approaches for the applications of LIF: the measurement method using fluorescent additives and the measurement of molecules that exist naturally in the measurement area. In medical applications,[8.19]–[8.23]

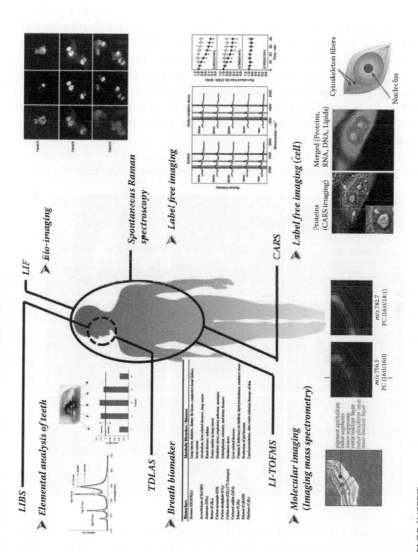

FIGURE 8.9 (SEE COLOR INSERT)

Summary of medical applications using LIF, LIBS, spontaneous Raman spectroscopy, CARS, TDLAS and LI-TOFMS. Many of the features of laser diagnostics, e.g., fast response, excellent special resolution, High sensitivity, *in situ* and noncontact detection, are also the major requirements of most of the medical applications. These applications extend from *in vivo* measurements of live cells to biological material analyses.

fluorescent additives are often used to measure a specific molecule in biological systems because there are numerous molecules within an organism and it is difficult to distinguish individual molecules from each other using LIF. The most prominent technique is the measurement method using fluorescent additives called "fluorescence probes." In this method fluorescently labeled proteins and antibodies are introduced into a live cell to see protein–protein interactions and kinetic changes. It is also possible to measure 3-D information using a confocal scanning laser microscope. These "labels" can be added to the individual protein and antibody to see each specific function in a live call. The drawback of fluorescence probes is that the fluorescently labeled proteins and antibodies do not always represent the movement of unleveled proteins and antibodies because labeled probes are part of the whole. This also occurs in the tracer-LIF methods in flow analyses: Tracers do not always show the same flow pattern as that of the target molecules.

Figure 8.10 shows the live-cell imaging results using LIF.[8.19] There are several methods and probes to visualize the cell structures, and they are

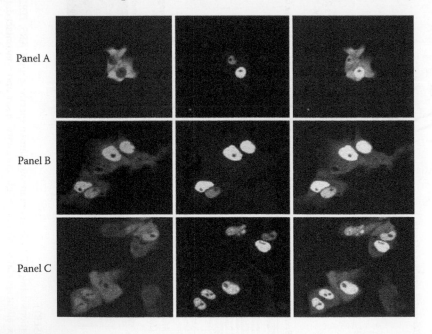

FIGURE 8.10
Live cell imaging using LIF. Confocal images of Cos-7 cells co-transfected with eCFP-AR and eYFPTif2. The first column corresponds to the eCFP channel, the second column to the eYFP channel and the third column is a merger of the two. Panel A represents cells imaged in absence of any ligand. Panel B corresponds to images of cells in presence of 10–8 M of the agonist R1881. Panel C corresponds to images of cells in presence of 10–8 M of the antagonist, casodex. (Reprinted with kind permission from **Springer Science+Business Media:** *European Biophysics Journal*, "Quantitative detection of the ligand-dependent interaction between the androgen receptor and the co-activator, Tif2, in live cells using two color, two photon fluorescence cross-correlation spectroscopy," 36(2), 2007, 153–161, T. Rosales et al.)

used for individual purposes. As spectroscopic methods, there have been mainly two types of excitation schemes to visualize fluorescence probes: one-photon (normal) fluorescence and two-photon fluorescence. Two-photon fluorescence has a better resolution in principle, and these methods have to be employed according to the measurement requirements.

8.4.2 Laser-Induced Breakdown Spectroscopy

As discussed in Chapter 4, LIBS is intrinsically an elemental analysis method; it employs a laser breakdown process to vaporize a small part of the measured materials to detect their atomic emissions. LIBS is not a nondestructive measurement technology in a precise sense, and its applications to medical fields have been developed differently from LIF or Raman spectroscopy. As for the solid material analyses, LIBS has a good spatial resolution, and 2-D elemental distribution on the surface of a material can be detected by scanning the laser focus point or the material position. Using this feature of LIBS, there have been several demonstrations to measure living materials, including bones and teeth.[8.24]–[8.28] Figure 8.11 shows the measurement results of Mg and Ca contents in a caries-infected tooth.[8.24] The differences in the spectra recorded for healthy and infected parts of the tooth are clear, and the ratio Mg/Ca can be an indicator to identify the infected parts of the tooth.

8.4.3 Spontaneous Raman Spectroscopy and CARS

LIF has been actively used to visualize the structure of a living cell and its kinetic changes using fluorescence probes. As with many industrial applications, there is a need for a direct measurement of naturally existing molecules in a living cell because it is not always possible to label biological molecules. As discussed in Chapter 5, the wavelength of Raman scattering light is shifted by energies corresponding to molecular ro-vibrational energy levels. The shift in energy gives information about molecules, called "molecular fingerprints." This feature of spontaneous Raman spectroscopy and CARS has been also employed to measure a living cell structure without labeling biological molecules. It is often called "label-free" analysis.[8.29]–[8.32]

Since spontaneous Raman spectroscopy is an easy and simple method both experimentally and theoretically, it is actively used to visualize kinetic changes of a living cell. Figure 8.12 shows the dependence of spontaneous Raman spectra in a living cell in different cell conditions.[8.29] After exposure to hydroxyl radicals for 40 minutes, the intensities of two bands decrease by 67% (1651 cm^{-1}) and 42% (1226 cm^{-1}), respectively. These results indicate that the C=C stretching and =CH modes disappear gradually during the oxidative process. In these applications, it is important and also challenging to relate the measurement results to the organic activities that form the complicated and delicate basis for the mechanisms of living organisms.

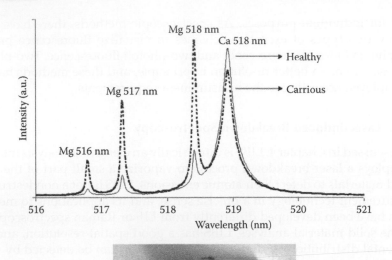

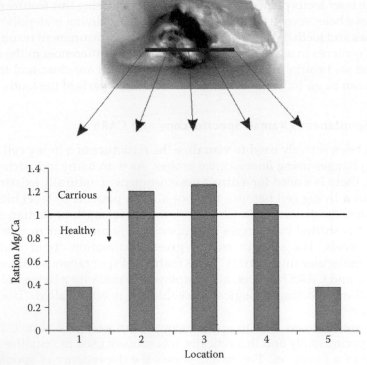

FIGURE 8.11

Elemental analysis of a tooth using LIBS. One-dimensional measurement map for the ratio of Mg content to Ca content, for a section of a caries-infected tooth. The increased Mg concentration clearly identifies the part of the tooth "softened" by caries. (Reprinted from *Spectrochimica Acta Part B*, 56(6), O. Samek et al., "Quantitative laser-induced breakdown spectroscopy analysis of calcified tissue samples," 865–875, 2001, with permission from Elsevier.)

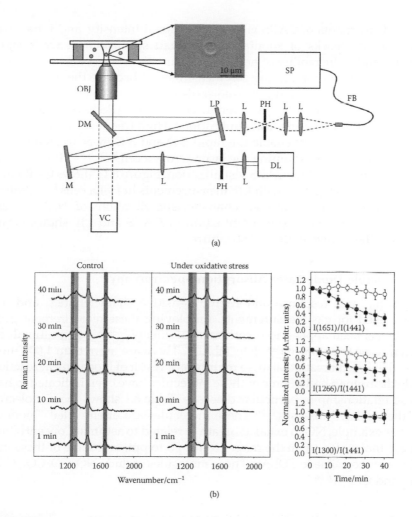

FIGURE 8.12
Spontaneous Raman spectroscopy for non-invasive biochemical analysis. (a) Experimental setup. The beam of a diode-pumped solid-state laser (DL) was expanded and collimated by a spatial filter that comprises two lenses (L) and a pinhole (PH). The beam was reflected by a mirror and a long-pass filter (LP), and was directed to a microscope objective (OBJ). The Raman signal collected by the same objective was transmitted through the long-pass filter and another spatial filter and was focused onto a fiber (FB) coupled spectrograph (SP). A video camera (VC) was installed to facilitate the observation of optically trapped single cells. The bright-field optical image captured by the video camera shows an optically trapped yeast in solution. (b) Effect of hydroxyl radical on single optically trapped yeasts. (a) Temporally varying Raman spectra of a single yeast cell. All spectra are background-subtracted and normalized with respect to the most intense 1441 cm^{-1} band, which shows little change over time to account for heterogeneity from cell to cell. For clarity, the two bands decreasing gradually under oxidative stress are highlighted in dark grey, whereas the other two that remained unaltered are highlighted in light grey. (b) Temporal evolution of relative intensities of Raman bands (empty circles: control, solid circles: single yeast cell under oxidative stress; n = 10). (From W.-T. Chang et al.: "Real-time molecular assessment on oxidative injury of single cells using Raman spectroscopy," *Journal of Raman Spectroscopy*, 2009, 40(9), 1194–1199. Copyright Wiley-VCH Verlag GmbH & Co. KGaA. Reproduced with permission.)

One of the merits of CARS is its strong signal intensity, and it has overcome the drawbacks of spontaneous Raman spectroscopy: weak signal intensity and vulnerability of noises, including fluorescence. These features are very important for a living cell measurement because there are numerous molecules in a cell, and these molecules are often found to be sources of fluorescence. Picosecond (ps) and femtosecond (fs) CARS have brought great advancement to CARS medical applications. High frequency and high peak intensity of those lasers enable the rapid 2-D and 3-D measurements of living cell structures by scanning the laser across a cell.[8.33]-[8.37] "Probes" have sometimes been used for CARS visualization. Figure 8.13 shows CARS measurement results together with LIF measurements in HeLa cells.[8.33] Proteins and lipids are observed at their characteristic vibrations of 2930 cm^{-1} and 2840 cm^{-1}, respectively. The combination of CARS and LIF shows clearer images of the change in the cell structure.

8.4.4 Tunable Diode Laser Absorption Spectroscopy

TDLAS has fast response and highly sensitive characteristics, and it is mainly used for gas measurements. Employing these measurement traits, TDLAS has been applied to breath measurements.[8.38]-[8.41] A multipath cell is often used to enhance the detectability. There are more than 1,000 molecules in exhaled breath that arise mainly as a result of biochemical reactions inside the human body. Some of these molecules have been indicated to have a close relationship to particular diseases. Table 8.1 shows these molecules and their related diseases.[8.38] These molecules are often called "biomarkers." For example, NO in exhaled breath is related to asthma. Concentrations of these molecules are in the range from ppb to ppt, and some of these molecules are accessible to TDLAS. These molecules include NO, CO, CO_2, C_2H_4, C_2H_6, and CH_2O.[8.38]

8.4.5 Laser Ionization Time-of-Flight Mass Spectrometry

TOFMS has been mainly developed in medical and pharmaceutical fields, and its main applications are still in these fields. Matrix-assisted laser desorption/ionization (MALDI)-TOFMS is one of the most famous methods that have been employed in medical and pharmaceutical applications. MALDI uses a "matrix," which is a molecule such as 3,5-dimethoxy-4-hydroxycinnamic acid, to reduce the fragmentation of measured molecules during the laser ionization. Biological molecules such as proteins and peptides have been analyzed using this method. In the medical applications of mass spectrometry, the 2-D measurement technology of imaging mass spectrometry (IMS) has been developed and applied to 2-D biomedical tissue measurements.[8.42]-[8.48] Figure 8.14 shows a typical setup of IMS. There are mainly

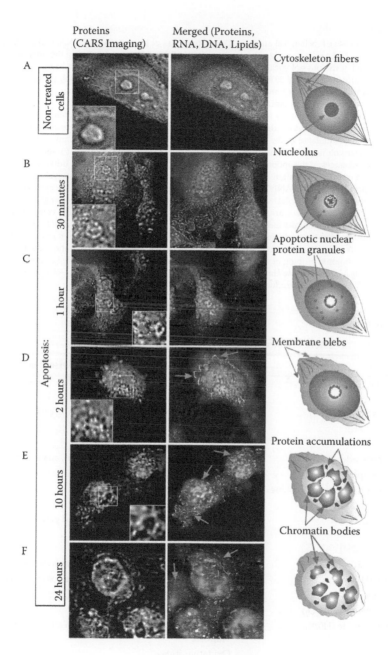

FIGURE 8.13
(Caption follows on next page)

FIGURE 8.13

CARS for non-invasive biochemical analysis. The distribution of proteins, lipids, DNA, and RNA in dividing and apoptotic HeLa cells visualized by multimodal CARS/TPEF imaging. During the imaging live cells were maintained at the physiological conditions. Proteins and lipids were observed in the CARS mode at their characteristic vibrations of 2930 cm^{-1} and 2840 cm^{-1}, respectively. Nucleic acids, stained by acridine orange, were acquired in the RNA and DNA fluorescence channels in TPEF mode. In the right panels, schematics of the macromolecular organization of cells are represented. The CARS signal from proteins is represented in the left panels. The panels in the middle represent merger signals of the proteins, RNA, DNA, and lipids. The white-outlined areas in the protein channel are enlarged below. (A) Nontreated cells. The signal from proteins is accumulated in the nucleolus (Inset, arrowhead) and the nuclear lamina (arrows). In the rest of the nuclear volume, the intensity of the protein signal is nearly uniform. (B–F) Representative cells at the subsequent stages of the apoptotic development. (B) 30 minutes following the initiation of apoptosis, the distribution of proteins is altered in the nucleolus (Inset, arrowheads) and the novel structure, apoptotic nuclear protein granules (ANPG), emerged in the nucleoplasm. In the cytoplasm, cytoskeleton fibers begin to lose their structural polarization. (C) The pattern of proteins becomes increasingly irregular, and ANPG become prominent (Inset, arrowheads). (D) The nucleolar proteins are forming a complex meshwork (Inset). The apoptotic membrane blebs are seen (D and E merged panels, arrows), ANPG disintegrate. (E and F) Proteins abandon the nucleolus and demonstrate a highly irregular distribution in the nucleoplasm; the genomic DNA is condensing to chromatin bodies and partially segregates from the proteins. (From A. Pliss et al.: "Biophotonic probing of macromolecular transformations during apoptosis," *Proceedings of the National Academy of Sciences of the United States of America*, 2010, 107(29), 12771–12776. Copyright Wiley-VCH Verlag GmbH & Co. KGaA. Reproduced with permission.)

TABLE 8.1

Biomarkers and Their Physiological Symptoms

Biomarkers	Metabolic Disorders/Diseases
Acetone (OC(CH$_3$)$_2$)	Lung cancer, diabetes, dietary fat losses, congestive heart failure, brain seizure
Acetaldehyde (CH$_3$CHO)	Alcoholism, liver-related diseases, lung cancer
Ammonia (NH$_3$)	Renal diseases, asthma
Butane (C$_4$H$_{10}$)	Tumor marker in lung cancer
Carbon monoxide (CO)	Oxidative stress, respiratory infection, anaemias
Carbon disulphide (CS$_2$)	Schizophrenia, coronary, and artery diseases
Carbon dioxide (CO$_2$) (^{13}C-Isotopes)	Oxidative stress
Carbonyl sulfide (OCS)	Liver-related diseases
Ethane (C$_2$H$_6$)	Vitamin E deficiency in children, lipid peroxidation, oxidative stress
Ethanol (C$_2$H$_5$OH)	Production of gut bacteria
Ethylene (C$_2$H$_4$)	Lipid peroxidation, ultraviolet radiation damage of skin
Hydrogen (H$_2$)	Indigestion in infants, intestinal upset, colonic fermentation
H/D isotope	Body water
Hydrogen peroxide (H$_2$O$_2$)	Asthma
Hydrogen cyanide (HCN)	Pseudomonas aeruginosa in children affected with cystic fibrosis
8-Isoprostane	Oxidative stress
Isoprene	Blood cholesterol
Methane (CH$_4$)	Intestinal problems, colonic fermentation
Methanethiol (CH$_3$SH)	Halitosis

(continued)

TABLE 8.1 (CONTINUED)

Biomarkers and Their Physiological Symptoms

Biomarkers	Metabolic Disorders/Diseases
Methanol (CH_3OH)	Nervous system disorder
Methylated amines	Protein metabolism in body
Methyl nitrate (CH_3NO_3)	Hyperglycemia in Type 1 diabetes
Nitrogen monoxide (NO)	Asthma, bronchiectasis, hypertension, rhinitis, lung diseases
Nitrotyrosine ($C_9H_{10}N_2O_5$)	Asthma
Oxygen (O_2)	Respiration
Pentane (C_5H_{12})	Peroxidation of lipids, liver diseases, schizophrenia, breast cancer, rheumatoid arthritis
Pyridine (C_5H_5N)	Periodontal disease
Sulfur compounds	Hepatic diseases and malordor, lung cancer
Hydrocarbons (Toluene ($C_6H_5CH_3$),	Lipid peroxidation, lung cancer, oxidative stress, airway inflammation
Benzene (C_6H_6), Heptane (C_7H_{16}),	
Decane ($C_{10}H_{22}$), Styrene (C_8H_8), Octane	
(C_8H_{18}), Pentamethylheptane ($C_{12}H_{26}$))	

Note that except for NO, the only one that has been approved by the U.S. Food and Drug Administration as a biomarker of chronic airway inflammation in asthma, other breath compounds listed in Table 1 should be more accurately termed as "potential biomarkers" while the term "biomarker(s)" is used throughout the paper.

Source: Reprinted from C. Wang and P. Sahay, "Breath analysis using laser spectroscopic techniques: Breath biomarkers, spectral fingerprints, and detection limits," *Sensors,* 9(10), 8230, 2009. With permission.

two devices in IMS: an apparatus for laser desorption and ionization of a biological tissue and a mass spectrometer. Two-dimensional distribution of mass spectra can be detected by scanning the laser desorption and ionization point on the tissue surface. It can be said that the method is a combination of LIBS and TOFMS. MARDI has often been used for a laser desorption and ionization process. Figure 8.15 shows one of the measurement results by

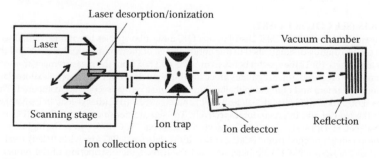

FIGURE 8.14
Typical setup of IMS. There are mainly two devices in IMS, i.e., an apparatus for laser desorption and ionization of a biological tissue and a mass spectrometer. 2-D distribution of mass spectra can be detected by scanning the laser desorption and ionization point on the tissue surface.

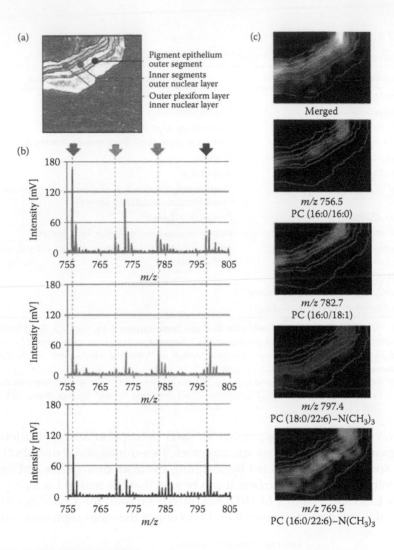

FIGURE 8.15 (SEE COLOR INSERT)
Typical measurement results of IMS. Distribution of PC molecular species in a mouse retinal section. (a) Each area can be roughly distinguished in this optical image of the mouse retinal section, and three colored dots are shown. (b) Three spectra between m/z 755 and 805 from the measurement areas at the red dot (outer nuclear layer and inner segment), green dot (inner nuclear layer and outer plexiform layer), and blue dot (outer segment and pigmentepithelium) in the mouse retinal section are compared. The signal intensities at m/z 756.5 ([PC (diacyl-16:0/16:0)+Na]+) and m/z 782.7 ([PC (diacyl-16:0/18:1) +Na]+) were the highest in the red dot and in green dot,respectively. The signal intensities at m/z 797.4 ([PC (diacyl-18-:0/22:6)+Na – N(CH3)3]+) and m/z 769 ([PC (diacyl-16:0/22:6)+Na – N(CH3)3]+) were the highest in the blue dot. (c) The ion image merged from [PC(diacyl-16:0/16:0)+Na]+ (red), [PC (diacyl-16:0/18:1) +Na]+ (green), and [PC (18:0/22:6)+Na – N(CH3)3]þ(blue) revealed the three-zone distribution of the retinal section. [PC (16:0/22:6)+Na – N(CH3)3]+ (light blue) was distributed in the same region as [PC (18:0/22:6)+Na – N(CH3)3]+. (From T. Hayasaka et al.: "Matrix-assisted laser desorption/ionization quadrupole ion trap time-of-flight (MALDI-QIT-TOF)-based imaging mass spectrometry reveals a layered distribution of phospholipid molecular species in the mouse retina," *Rapid Communications in Mass Spectrometry*," 2008, 22(21), 3415–3426. Copyright Wiley-VCH Verlag GmbH & Co. KGaA. Reproduced with permission.)

IMS.[8.48] The distributions of molecular species at a mouse retinal section are clearly shown using IMS.

References

[8.1]   B.E. Bouma, S.-H. Yun, B.J. Vakoc, M. J. Suter, and G.J. Tearney, "Fourier-domain optical coherence tomography: Recent advances toward clinical utility," *Current Opinion in Biotechnology*, 20(1), 111–118, 2009.

[8.2]   M.G. Sowa, D.P. Popescu, J. Werner, M. Hewko, A.C.-T. Ko, J. Payette, C.C.S. Dong, B. Cleghorn, and L.-P. Choo-Smith, "Precision of Raman depolarization and optical attenuation measurements of sound tooth enamel," *Analytical and Bioanalytical Chemistry*, 387(5), 1613–1619, 2007.

[8.3]   J. Na, H.J. Baek, S.Y. Ryu, C. Lee, and B.H. Lee, "Tomographic imaging of incipient dental-caries using optical coherence tomography and comparison with various modalities," *Optical Review*, 16(4), 426–431, 2009.

[8.4]   M. Mogensen, L. Thrane, T.M. Joergensen, P.E. Andersen, and G.B.E. Jemec, "Optical coherence tomography for imaging of skin and skin diseases," *Seminars in Cutaneous Medicine and Surgery*, 28(3), 196–202, 2009.

[8.5]   M. Bonesi, S. Matcher, and I. Meglinski, "Doppler optical coherence tomography in cardiovascular applications," *Laser Physics*, 20(6), 1491–1499, 2010.

[8.6]   K.V. Larin, M.G. Ghosn, S.N. Ivers, A. Tellez, and J.F. Granada, "Quantification of glucose diffusion in arterial tissues by using optical coherence tomography," *Laser Physics Letters*, 4(4), 312–317, 2007.

[8.7]   K.A. Serrels, M.K. Renner, and D.T. Reid, "Optical coherence tomography for non-destructive investigation of silicon integrated-circuits," *Microelectronic Engineering*, 87(9), 1785–1791, 2010.

[8.8]   T. Prykaeri, J. Czajkowski, E. Alarousu, and R. Myllylae, "Optical coherence tomography as an accurate inspection and quality evaluation technique in paper industry," *Optical Review*, 17(3), 218–222, 2010.

[8.9]   M. Wiesner, J. Ihlemann, H.H. Muller, E. Lankenau, and G. Huttmann, "Optical coherence tomography for process control of laser micromachining," *Review of Scientific Instruments*, 81(3), 033705, 2010.

[8.10] H. Harde, J. Zhao, M. Wolff, R.A. Cheville, and D. Grischkowsky, "THz time-domain spectroscopy on ammonia," *Journal of Physical Chemistry A*, 105(25), 6038–6047, 2001.

[8.11] V.B. Podobedov, D.F. Plusquellic, K.E. Siegrist, G.T. Fraser, Q. Ma, and R.H. Tipping, "New measurements of the water vapor continuum in the region from 0.3 to 2.7THz," *Journal of Quantitative Spectroscopy and Radiative Transfer*, 109(3), 458–467, 2007.

[8.12] M.R. Leahy-Hoppa, J. Miragliotta, R. Osiander, J. Burnett, Y. Dikmelik, C. McEnnis, and J.B. Spicer, "Ultrafast laser-based spectroscopy and sensing: Applications in LIBS, CARS, and THz spectroscopy," *Sensors*, 10, 4342–4372, 2010.

[8.13] M.R. Leahy-Hoppa, M.J. Fitch, X. Zheng, L.M. Hayden, and R. Osiander, "Wideband terahertz spectroscopy of explosives," *Chemical Physics Letters*, 434(4–6), 227–230, 2007.

[8.14] H. Wu, E.J. Heilweil, A.S. Hussain, and M.A. Khan, "Process analytical technology (PAT): Quantification approaches in terahertz spectroscopy for pharmaceutical application," *Journal of Pharmaceutical Sciences*, 97(2), 970–984, 2007.

[8.15] P.C. Ashworth, E. Pickwell-MacPherson, E. Provenzano, S.E. Pinder, A.D. Purushotham, M. Pepper, and V.P. Wallace, "Terahertz pulsed spectroscopy of freshly excised human breast cancer," *Optics Express*, 17(15), 12444–12454, 2009.

[8.16] J. Nishizawa, T. Sasaki, K. Suto, T. Yamada, T. Tanabe, T. Tanno, T. Sawai, and Y. Miura, "THz imaging of nucleobases and cancerous tissue using a GaP THz-wave generator," *Optics Communications*, 244(1–6), 469–474, 2005.

[8.17] K. Kawase, T. Shibuyaa, S. Hayash, and K. Suizu, "THz imaging techniques for nondestructive inspections," *Comptes Rendus Physique*, 11(7–8), 510–518, 2010.

[8.18] F.F. Sizov, "THz radiation sensors," *Opto-Electronics Review*, 18(1), 10–36, 2010.

[8.19] T. Rosales, V. Georget, D. Malide, A. Smirnov, J. Xu, C. Combs, J.R. Knutson, J.-C. Nicolas, and C.A. Royer, "Quantitative detection of the ligand-dependent interaction between the androgen receptor and the co-activator, Tif2, in live cells using two color, two photon fluorescence cross-correlation spectroscopy," *European Biophysics Journal*, 36(2), 153–161, 2007.

[8.20] M. Baker, "Cellular imaging: Taking a long, hard look," *Nature*, 466(7310), 1137–1140, 2010.

[8.21] H.M. Kim, B.R. Kim, H.-J. Choo, Y.-G. Ko, S.-J. Jeon, C.H. Kim, J. Taiha, and B.R. Cho, "Two-photon fluorescent probes for biomembrane imaging: Effect of chain length," *ChemBioChem*, 9(17), 2830–2838, 2008.

[8.22] P.S. Mohan, C.S. Lim, Y.S. Tian, W.Y. Roh, J.H. Lee, and B.R. Cho, "A two-photon fluorescent probe for near-membrane calcium ions in live cells and tissues," *Chemical Communications*, 5365–5367, 2009.

[8.23] V.P. Tokar, Y.L. Mykhaylo, Y. Tymish, V.K. Dmytro, B.K. Vladyslava, O.B. Anatoliy, M.D. Igor, M.P. Vadym, M.Y. Sergiy, and M.Y. Valeriy, "Styryl dyes as two-photon excited fluorescent probes for DNA detection and two-photon laser scanning fluorescence microscopy of living cells," *Journal of Fluorescence*, 20(4), 865–872, 2010.

[8.24] O. Samek, D.C.S. Beddowsb, H.H. Telle, J. Kaiser, M. Liska, J.O. Caceres, and A.G. Urena, "Quantitative laser-induced breakdown spectroscopy analysis of calcified tissue samples," *Spectrochimica Acta Part B*, 56(6), 865–875, 2001.

[8.25] M. Galiová, J. Kaiser, F.J. Fortes, K. Novotný, R. Malina, L. Prokeš, A. Hrdlička, T. Vaculovič, M.N. Fišáková, J. Svoboda, V. Kanický, and J J. Laserna, "Multielemental analysis of prehistoric animal teeth by laser-induced breakdown spectroscopy and laser ablation inductively coupled plasma mass spectrometry," *Applied Optics*, 49(13), C191–C199, 2010.

[8.26] F.C. Alvira, F.R. Rozzi, and G.M. Bilmes, "Laser-induced breakdown spectroscopy microanalysis of trace elements in Homo sapiens teeth," *Applied Spectroscopy*, 64(3), 313–319, 2010.

[8.27] V.K. Singh and A.K. Rai, "Potential of laser-induced breakdown spectroscopy for the rapid identification of carious teeth," *Lasers in Medical Science*, 26(3), 307–315, 2011.

[8.28] R.K. Thareja, A.K. Sharma, and S. Shukla, "Spectroscopic investigations of carious tooth decay," *Medical Engineering and Physics*, 30(9), 1143–1148, 2008.

[8.29] W.-T. Chang, H.-L. Lin, H.-C. Chen, Y.-M. Wu, W.-J. Chen, Y.-T. Lee, and I. Liau, "Real-time molecular assessment on oxidative injury of single cells using Raman spectroscopy," *Journal of Raman Spectroscopy*, 40(9), 1194–1199, 2009.

[8.30] R.J. Swain and M.M. Stevens, "Raman microspectroscopy for non-invasive bio-chemical analysis of single cells," *Biochemical Society Transactions*, 35(3), 544–549, 2007.

[8.31] T. Huser, C.A. Orme, C.W. Hollars, M.H. Corzett, and R. Balhorn, "Raman spectroscopy of DNA packaging in individual human sperm cells distinguishes normal from abnormal cells," *Journal of Biophotonics*, 2(5), 322–332, 2009.

[8.32] W.E. Huang, M.J. Bailey, I.P. Thompson, A.S. Whiteley, and A.J. Spiers, "Single-cell Raman spectral profiles of pseudomonas fluorescens SBW25 reflects in vitro and in planta metabolic history," *Microbial Ecology*, 53(3), 414–425, 2007.

[8.33] A. Pliss, A.N. Kuzmin, A.V. Kachynski, and P.N. Prasad, "Biophotonic probing of macromolecular transformations during apoptosis," *Proceedings of the National Academy of Sciences of the United States of America*, 107(29), 12771–12776, 2010.

[8.34] R.S. Lim, A. Kratzer, N.P. Barry, S. Miyazaki-Anzai, M. Miyazaki, W.W. Mantulin, M. Levi, E.O. Potma, and B.J. Tromberg, "Multimodal CARS microscopy determination of the impact of diet on macrophage infiltration and lipid accumulation on plaque formation in ApoE-deficient mice," *Journal of Lipid Research*, 51(7), 1729–1737, 2010.

[8.35] S.-H. Kim, E.-S. Lee, J.Y. Lee, E.S. Lee, B.-S. Lee, J.E. Park, and D.W. Moon, "Multiplex coherent anti-Stokes Raman spectroscopy images intact atheromatous lesions and concomitantly identifies distinct chemical profiles of atherosclerotic lipids," *Circulation Research*, 106(8), 1332–1341, 2010.

[8.36] X. Nan, E.O. Potma, and X.S. Xie, "Nonperturbative chemical imaging of organelle transport in living cells with coherent anti-Stokes Raman scattering microscopy," *Biophysical Journal*, 91(2), 728–735, 2006.

[8.37] R. Mouras, G. Rischitor, A. Downes, D. Salterb, and A. Elfick, "Nonlinear optical microscopy for drug delivery monitoring and cancer tissue imaging," *Journal of Raman Spectroscopy*, 41(8), 848–852, 2010.

[8.38] C. Wang and P. Sahay, "Breath analysis using laser spectroscopic techniques: Breath biomarkers, spectral fingerprints, and detection limits," *Sensors*, 9(10), 8230–8262, 2009.

[8.39] M. Phillips, R.N. Cataneo, T. Cheema, and J. Greenberg, "Increased breath biomarkers of oxidative stress in diabetes mellitus,"*Clinica Chimica Acta*, 344(1–2), 189–194, 2004.

[8.40] K.R. Parameswaran, D.I. Rosen, M.G. Allen, A.M. Ganz, and T.H. Risby, "Off-axis integrated cavity output spectroscopy with a mid-infrared interband cascade laser for real-time breath ethane measurements," *Applied Optics*, 48(4), B73–B79, 2009.

[8.41] S. Engel, H.M. Lease, N.G. McDowell, A.H. Corbett, and B.O. Wolf, "The use of tunable diode laser absorption spectroscopy for rapid measurements of the δ13C of animal breath for physiological and ecological studies," *Rapid Communications in Mass Spectrometry*, 23(9), 1281–1286, 2009.

[8.42] R.M.A. Heeren, D.F. Smith, J. Stauber, B. Kükrer-Kaletas, and L. MacAleese, "Imaging mass spectrometry: Hype or hope?" *Journal of the American Society for Mass Spectrometry*, 20(6), 1006–1014, 2009.

[8.43] C. Murayama, Y. Kimura, and M. Setou, "Imaging mass spectrometry: Principle and application," *Biophysical Reviews*, 1(3), 131–139, 2009.

[8.44] K. Chughtai and R.M.A. Heeren, "Mass spectrometric imaging for biomedical tissue analysis," *Chemical Reviews*, 110(5), 3237–3277, 2010.

[8.45] S. Koizumi, S. Yamamoto, T. Hayasaka, Y. Konishi, M. Yamaguchi-Okada, N. Goto-Inoue, Y. Sugiura, M. Setou, and H. Namba, "Imaging mass spectrometry revealed the production of lyso-phosphatidylcholine in the injured ischemic rat brain," *Neuroscience*, 168(1), 219–225, 2010.

[8.46] D.H. Hölscher, R. Shroff, K. Knop, M. Gottschaldt, A. Crecelius, B. Schneider, D.G. Heckel, U.S. Schuber, and A. Svatos, "Matrix-free UV-laser desorption/ionization (LDI) mass spectrometric imaging at the single-cell level: Distribution of secondary metabolites of Arabidopsis thaliana and Hypericum species," *Plant Journal*, 60(5), 907–918, 2009.

[8.47] E.H. Seeley and R.M. Caprioli, "Imaging mass spectrometry: Towards clinical diagnostics," *Proteomics: Clinical Applications*, 2(10–11), 1435–1443, 2008.

[8.48] T. Hayasaka, N. Goto-Inoue, Y. Sugiura, N. Zaima, H. Nakanishi, K. Ohishi, S. Nakanishi, T. Naito, R. Taguchi, and M. Setou, "Matrix-assisted laser desorption/ionization quadrupole ion trap time-of-flight (MALDI-QIT-TOF)-based imaging mass spectrometry reveals a layered distribution of phospholipid molecular species in the mouse retina," *Rapid Communications in Mass Spectrometry*, 22(21), 3415–3426, 2008.

Appendix A: Summary of Laser Diagnostics

TABLE A.1

Laser-Induced Fluorescence

Principle	

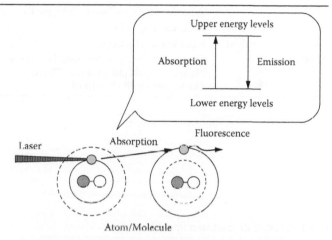

Following the absorption of this incident light, the molecule undergoes emission, collision, and other processes to transfer into other energy states. The emission is known as fluorescence and is used to determine concentration and temperature.

Geometric arrangement	

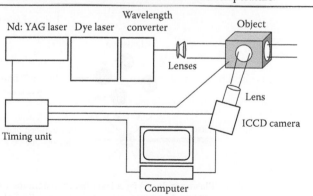

Main components of a LIF system are a laser and a CCD camera. An ICCD camera is often used to pick out the fast LIF signal from other noise signals.

(continued)

TABLE A.1 (CONTINUED)

Laser-Induced Fluorescence

Item	• Species concentration OH, NO, O_2, CH, etc. • Tracers (biacety, acetone, formaldehyde, etc.) • Temperature • Velocity • Pressure
Detection limit	ppm–%
Laser	Tunable lasers (dye lasers, OPO, etc.)
Detector	ICCD camera + filter
Conventional method	Emission spectroscopy
Applications	Combustion analyses (engines, burners, gas turbines, etc.) Plasma analysis (arc plasmas, CVDs) Life sciences (cell visualization)

TABLE A.2

Laser Induced Breakdown Spectroscopy

Principle	

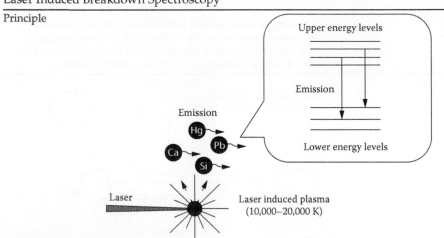

Plasma induced by a laser beam instantaneously reaches the 10,000–20,000K temperature regions. The plasma first emits strong continuous noise, with atomic emissions appearing after a certain time delay. The necessary time delay is dependent on the upper energy of the measured element and plasma conditions.

(continued)

TABLE A.2 (CONTINUED)

Laser Induced Breakdown Spectroscopy

Geometric arrangement	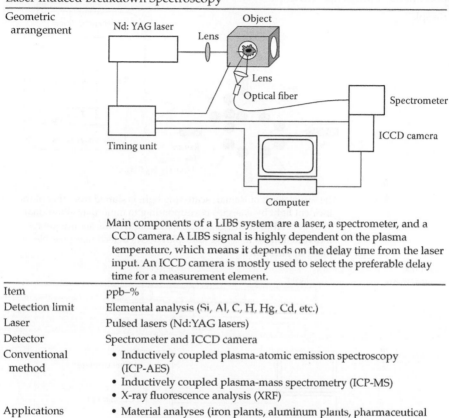
	Main components of a LIBS system are a laser, a spectrometer, and a CCD camera. A LIBS signal is highly dependent on the plasma temperature, which means it depends on the delay time from the laser input. An ICCD camera is mostly used to select the preferable delay time for a measurement element.
Item	ppb–%
Detection limit	Elemental analysis (Si, Al, C, H, Hg, Cd, etc.)
Laser	Pulsed lasers (Nd:YAG lasers)
Detector	Spectrometer and ICCD camera
Conventional method	• Inductively coupled plasma-atomic emission spectroscopy (ICP-AES) • Inductively coupled plasma-mass spectrometry (ICP-MS) • X-ray fluorescence analysis (XRF)
Applications	• Material analyses (iron plants, aluminum plants, pharmaceutical plants, thermal power plants, etc.) • Safety and security (heavy metals, explosives, etc.)

TABLE A.3

Spontaneous Raman Spectroscopy

Principle	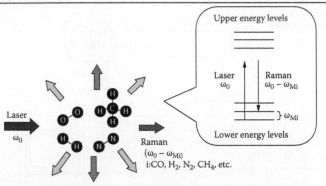

The wavelength of Raman scattering light is shifted from that of the incident light by energies corresponding to molecular vibrational and/or rotational energies. The wavelength shifts are unique for individual molecules and multiple species detection is possible in many applications.

Geometric arrangement	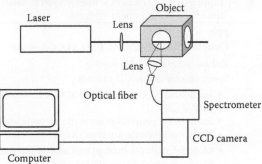

Main components of a Raman spectroscopy system are a laser, a spectrometer, and a CCD camera. A Raman scattering signal is very weak and sensitive detectors are needed to get sufficient signals.

Item	• Species concentration (major species)
	• N_2, O_2, H_2, H_2O, CO_2, CH_4, etc.
	• Temperature
Detection limit	%
Laser	• Pulsed lasers (Nd:Yag lasers),
	• Continuous lasers (diode lasers)
Detector	• Spectrometer + ICCD camera
	• Spectrometer + EMCCD camera
Conventional method	• Concentration : Gas
	• Temperature : Thermocouple
Applications	• Combustion analyses (engines, burners, gas turbines, etc.)
	• Process monitoring [pharmaceutical plants, semiconductor plants (CVD), etc.]
	• Life sciences (label-free cell visualization)

TABLE A.4

CARS

Principle	

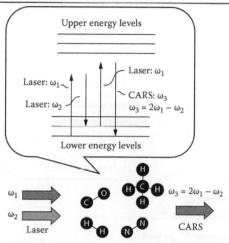

CARS is one of the nonlinear optical processes. Using a pump beam ω_1 and a probe beam ω_2, the CARS signal with $\omega_3 = 2\omega_1 - \omega_2$ is generated by Raman effects.

Geometric arrangement

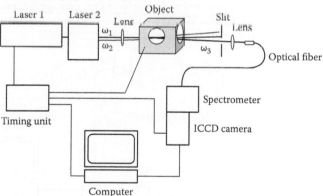

In CARS two laser systems are usually employed for pump and probe beams. They are crossed with each other and a beamlike CARS signal is generated in the phase-matching condition.

Item	• Species concentration (major species)
	N_2, O_2, H_2, H_2O, CO_2, CH_4, etc.
	• Temperature
Detection limit	%
Laser	• Pulsed lasers (Nd:Yag lasers) + tunable lasers (dye lasers)
Detector	• Spectrometer + ICCD camera
Conventional method	• Concentration : Gas
	• Temperature : Thermocouple
Applications	• Combustion analyses (engines, burners, gas turbines, etc.)
	• Life sciences (label-free cell visualization)

TABLE A.5

Tunable Diode Laser Absorption Spectroscopy

Principle	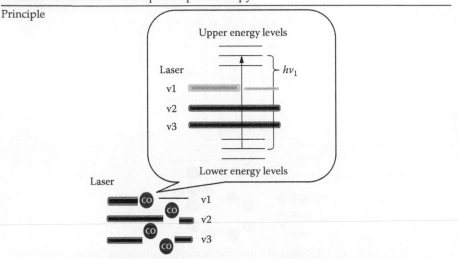

When light permeates an absorption medium, the strength of the permeated light is related to absorber concentration according to Lambert Beer's law. Atoms and molecules have their own spectral patterns. Because of this feature, TDLAS has excellent selectivity and sensitivity.

Geometric arrangement	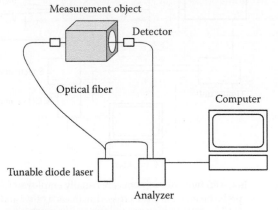

The TDLAS system is simple, and its main components are a diode laser and a detector. Modulation methods or balanced receivers are often used to get small absorption signals.

Item	• Species concentration
	CO, CO_2, H_2O, CO_2, CH_4, NO, N_2O, etc.
	• Temperature
Detection limit	• ppb–%
Laser	• Diode lasers
Detector	• Photo-diodes

(continued)

TABLE A.5 (CONTINUED)

Tunable Diode Laser Absorption Spectroscopy

Conventional method	• Concentration: Gas chromatography (GC) Fourier transform infrared spectroscopy (FTIR), Nondispersive infrared spectroscopy (NDIR) • Temperature: Thermocouple
Applications	• Combustion analyses (engines, burners, gas turbines, etc.) • Process monitoring (thermal power plants, disposal plants, pharmaceutical plants, etc.)

TABLE A.6

Laser Ionization Time-of-Flight Mass Spectrometry

Principle	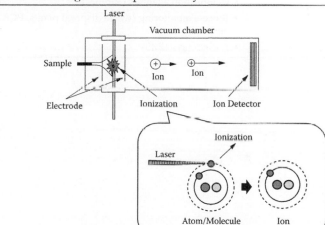 Unlike other laser diagnostics, a laser ionization TOFMS method samples the gas to a vacuum chamber. In the TOFMS process, ions can be counted by the ion detector; therefore, super-high sensitivity can be achieved using this method.
Geometric arrangement	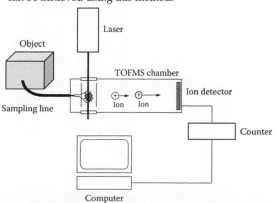

(*continued*)

TABLE A.6 (CONTINUED)

Laser Ionization Time-of-Flight Mass Spectrometry

	The main components are a laser and a TOFMS chamber. Supersonic jet cooling and ion trap methods are often used to enhance the sensitivity
Item	• Species concentration
	Benzene, PAHs, PCBs, DXNs, etc.
Detection limit	• ppt
Laser	• Pulsed lasers (Nd:Yag lasers)
	• Tunable lasers (dye lasers, OPOs)
Detector	• Ion detector (MCPs)
Conventional method	• Gas chromatography–mass spectrometry (GC-MS)
	• Gas chromatography–electron capture detector(GC-ECD)
Applications	• Combustion analyses (engines)
	• Process monitoring (waste disposal plants, PCB disposal plants, etc.)
	• Life sciences (IMS)

Appendix B: Molecular Energy States and Energy Transfers

Internal energy of a molecular system is given as a sum of electronic, vibrational, and rotational energies. These energies are given as eigenvalues of each eigenfunction ψ_e, ψ_v, and ψ_r of the Schrödinger equation in the molecular system. The total energy is given by the sum of the electronic, vibrational, and vibrational energies

$$E = E_e + E_v + E_r \tag{B.1}$$

In diatomic molecules, vibrational and rotational energies are quantized according to the relations

$$E_v = hc\left(\omega_e\left(v+\frac{1}{2}\right) - \omega_e x_e\left(v+\frac{1}{2}\right)^2 + \omega_e y_e\left(v+\frac{1}{2}\right)^3 \cdots\right) \tag{B.2}$$

$$F_r = hc\left(B_v J(J+1) - D_v J^2(J+1)^2 + \cdots\right) \tag{B.3}$$

where h is the Plank's constant, c the speed of light, v the vibrational quantum number, and J the rotational quantum number. ω_e and $\omega_e x_e$ are the vibrational constants, and B_v and D_v are the rotational constants. Equations (B.2) and (B.3) are found by solution of the simplified Schrödinger equation. This concept is also applicable to polyatomic molecules. B_v and D_v are modeled by the following relations:

$$B_v = B_e - \alpha_e\left(v+\frac{1}{2}\right) \tag{B.4}$$

$$D_v = D_e + \beta_e\left(v+\frac{1}{2}\right) \tag{B.5}$$

Here B_e and D_e are the rotational constants at the equilibrium position, and α_e and β_e are the vibration-rotation interaction constants.

In the molecular energy transitions, the selection rules stipulate the possible transitions. The selection rules depend on the characteristics of molecular energy states such as symmetries, nuclear spins, quantum numbers, and so on. It is useful to remember "names" of the rotational transfer notation.

These are five "branches" according to the $\Delta J = J' - J$ in $(v, J) \rightarrow (v', J')$ energy transitions:

O branch : $\Delta J = -2$
P branch : $\Delta J = -1$
Q branch : $\Delta J = 0$
R branch : $\Delta J = +1$
S branch : $\Delta J = +2$

These names have some variations such as Q_1, Q_2, and Q_{21} according to electric energy states of molecules. It is often enough to check them in tables of related papers or databases including HITRAN.

Appendix C: Line Broadenings

There are three types of line broadenings: natural, Doppler, and collision. The natural broadening occurs according to the Heisenberg uncertainty principle. It relates the lifetime of an excited state with the uncertainty of its energy. The natural broadening can be given by

$$G_N(v) = \frac{v_N}{2\pi} \frac{1}{(v-v_0)^2 + \left(\frac{v_N}{2}\right)^2} \tag{C.1}$$

where Δv_N is FWHM of the natural broadening. Usually the natural broadening is small and does not have considerable contribution to actual spectra observed in practical applications. The Doppler and collision broadenings are dominant in real applications and have the line shape functions

$$G_D(v) = \frac{c}{v_0}\sqrt{\frac{m}{2\pi kT}} \exp\left[-4\ln 2 \cdot \frac{(v-v_0)^2}{v_D^2}\right] \tag{C.2}$$

$$G_C(v) = \frac{v_C}{2\pi} \frac{1}{(v-v_0)^2 + \left(\frac{v_C}{2}\right)^2} \tag{C.3}$$

where $G_D(v)$ and $G_C(v)$ are the Doppler and collision broadenings respectively, v is the frequency of light, v_0 the transition center frequency, c the speed of light, m the atomic or molecular mass, k the Boltzmann constant, and Δv the transition full width at half maximum (FWHM). Δv_D and Δv_C are FWHMs of the Doppler and collision broadenings. Here, Δv_D are given by

$$v_D = \frac{2v_0}{c}\sqrt{\frac{2\ln 2 \cdot kT}{m}} \tag{C.4}$$

There is also spectral line shifting according to the collision effects. The combination of Doppler and collision broadenings is described by the Voigt function

$$G_V(a,x) = \frac{a}{\pi} \int_{-\infty}^{\infty} \frac{e^{-y^2}}{a^2 + (x-y)^2} dy \tag{C.6}$$

a and x are defined as

$$a = \sqrt{\ln 2}\, \frac{v_C}{v_D} \tag{C.7}$$

$$x = \sqrt{\ln 2}\, \frac{(v-v_0)}{v_D} \tag{C.8}$$

271

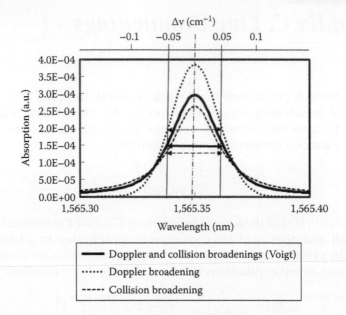

FIGURE C.1
Line broadenings. The Doppler and collision broadenings are dominant in practical applications. The combination of the Doppler and collision broadenings is described by the Voigt function. The collision broadening is dominant at high pressure.

These broadening shapes are shown in Figure C.1. The line shape functions are normalized to form the following relationship:

$$\int G(v)dv = 1 \tag{C.9}$$

The line shape of absorption spectra is important to evaluate the quantitative measurement. The Voigt function is easily calculated and often used for modeling of the absorption and emission spectra.

Appendix D: Boltzmann Distribution

There are rotational, vibrational, and electronic energy states in molecules, and these energy structures are related to the number density of molecules at each energy state. The number density n_i of a molecule at the energy state i at temperature T is described by the Boltzmann equation

$$n_i = n \frac{g_i e^{-E_i/kT}}{\sum_{i'} g_{i'} e^{-E_{i'}/kT}}$$

$$= n \frac{g_i e^{-E_i/kT}}{Z} \tag{D.1}$$

where g_i and E_i are the degeneracy and the energy of i state, k the Boltzmann constant, and Z the partition function. The total energy E is given by the sum of the electric, vibrational, and vibrational energies (see also Appendix B).

$$E = E_e + E_v + E_r \tag{D.2}$$

where E_e, E_v, and E_r are the electronic, vibrational, and rotational energies, respectively. In many applications it is often enough to consider the number density of molecules at rotational and vibrational energy states. The number density of molecules at a (v,J) ro-vibrational state is given by

$$n_{v,j} = n \frac{g_{v,j} e^{-E_{v,j}/kT}}{\sum_{v,j} g_{v,j} e^{-E_{v,j}/kT}}$$

$$= n \frac{g_{v,j} e^{-E_{v,j}/kT}}{Z} \tag{D.3}$$

where $g_{v,J}$ and $E_{v,J}$ are the degeneracy and the energy of the (v,J) state, respectively. The partition function Z can be given by the product of vibrational and rotational partition functions

$$Z = Z_v \cdot Z_r \tag{D.4}$$

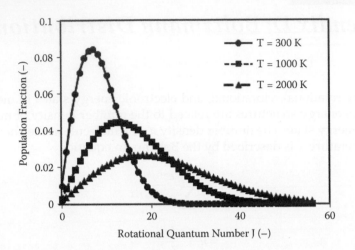

FIGURE D.1

Boltzmann distributions at different temperature conditions. As the temperature becomes lower, molecules exist only at energy levels with small rotational quantum numbers. The rotational constant $B = 2$ cm^{-1} and vibrational constant $\omega_e = 2300$ cm^{-1} are used to evaluate the Boltzmann distributions.

Here Z_v and Z_r are the vibrational and rotational partition functions and are given by the following abbreviated equations (see Appendix B):

$$Z_v = \sum_v g_v e^{-E_v/kT}$$

$$= \sum_v e^{-hc\omega_e v/kT}$$

$$= \frac{1}{1 - e^{-hc\omega_e/kT}} \tag{D.5}$$

$$Z_r = \sum_J g_J e^{-E_J/kT}$$

$$= \sum_J (2J+1)e^{-hcBJ(J+1)/kT}$$

$$= \frac{kT}{hcB} \tag{D.6}$$

Here h is the Planck's constant, c the speed of light, v the vibrational quantum number, and J the rotational quantum number. ω_e is the vibrational constant and B the rotational constant.

In the vibrational partition function Z_v, the zero-point energy at $v = 0$ can be neglected since it divides out in Equation (D.3). Equations (D.5) and (D.6) are the abbreviated methods to evaluate the partition function, and the results often show reasonable results for the evaluation of the number densities of molecules at ro-vibrational states. Figure D.1 shows an example of Boltzmann distributions at different temperature conditions.

Index

Printed and bound by CPI Group (UK) Ltd, Croydon, CR0 4YY

23/10/2024

01778227-0012